Reading Essen[tials]
An Interactive Student Textbook

Focus On
Earth Science

CALIFORNIA
GRADE
6

ca6.msscience.com

Mc Graw Hill **Glencoe**

New York, New York   Columbus, Ohio   Chicago, Illinois   Peoria, Illinois   Woodland Hills, California

To the Student

In today's world, knowing science is important for thinking critically, solving problems, and making decisions. But understanding science sometimes can be a challenge.

Reading Essentials takes the stress out of reading, learning, and understanding science. This book covers important concepts in science, offers ideas for how to learn the information, and helps you review what you have learned.

In each chapter:

- **Before You Read** sparks your interest in what you'll learn and relates it to your world.
- **Read to Learn** describes important science concepts with words and graphics. Next to the text you can find a variety of study tips and ideas for organizing and learning information:
  - The **Study Coach** offers tips for getting the main ideas out of the text.
  - **Foldables™ Study Organizers** help you divide the information into smaller, easier-to-remember concepts.
  - **Reading Checks** ask questions about key concepts. The questions are placed so you know whether you understand the material.
  - **Think It Over** elements help you consider the material in-depth, giving you an opportunity to use your critical-thinking skills.
  - **Picture This** questions specifically relate to the art and graphics used with the text. You'll find questions to get you actively involved in illustrating the concepts you read about.
  - **Applying Math** reinforces the connection between math and science.
  - **Academic Vocabulary** defines some important words that will help you build a strong vocabulary.

The main California Science Content Standard for a lesson appears at the beginning of each lesson. This statement explains the essentials skills and knowledge that you will be building as you read the lesson. A complete listing of the **Grade Six Science Content Standards** appears on pages iv to vi.

See for yourself, *Reading Essentials* makes science enjoyable and easy to understand.

Glencoe

The McGraw-Hill Companies

Send all inquiries to:
Glencoe/McGraw-Hill
8787 Orion Place
Columbus, OH 43240-4027

ISBN-13: 978-0-07-879430-8
ISBN-10: 0-07-879430-7
Printed in the United States of America
4 5 6 7 8 9 10 047 11 10 09 08

Table of Contents

Grade 6 Science Content Standards

1. **Plate tectonics accounts for important features of Earth's surface and major geologic events. As a basis for understanding this concept:**

   a. Students know evidence of plate tectonics is derived from the fit of the continents; the location of earthquakes, volcanoes, and midocean ridges; and the distribution of fossils, rock types, and ancient climatic zones.

   b. Students know Earth is composed of several layers: a cold, brittle lithosphere; a hot, convecting mantle; and a dense, metallic core.

   c. Students know lithospheric plates the size of continents and oceans move at rates of centimeters per year in response to movements in the mantle.

   d. Students know that earthquakes are sudden motions along breaks in the crust called faults and that volcanoes and fissures are locations where magma reaches the surface.

   e. Students know major geologic events, such as earthquakes, volcanic eruptions, and mountain building, result from plate motions.

   f. Students know how to explain major features of California geology (including mountains, faults, volcanoes) in terms of plate tectonics.

   g. Students know how to determine the epicenter of an earthquake and know that the effects of an earthquake on any region vary, depending on the size of the earthquake, the distance of the region from the epicenter, the local geology, and the type of construction in the region.

2. **Topography is reshaped by the weathering of rock and soil and by the transportation and deposition of sediment. As a basis for understanding this concept:**

   a. Students know water running downhill is the dominant process in shaping the landscape, including California's landscape.

   b. Students know rivers and streams are dynamic systems that erode, transport sediment, change course, and flood their banks in natural and recurring patterns.

   c. Students know beaches are dynamic systems in which the sand is supplied by rivers and moved along the coast by the action of waves.

   d. Students know earthquakes, volcanic eruptions, landslides, and floods change human and wildlife habitats.

3. **Heat moves in a predictable flow from warmer objects to cooler objects until all the objects are at the same temperature. As a basis for understanding this concept:**

   a. Students know energy can be carried from one place to another by heat flow or by waves, including water, light and sound waves, or by moving objects.

   b. Students know that when fuel is consumed, most of the energy released becomes heat energy.

   c. Students know heat flows in solids by conduction (which involves no flow of matter) and in fluids by conduction and by convection (which involves flow of matter).

   d. Students know heat energy is also transferred between objects by radiation (radiation can travel through space).

4. **Many phenomena on Earth's surface are affected by the transfer of energy through radiation and convection currents. As a basis for understanding this concept:**

   a. Students know the sun is the major source of energy for phenomena on Earth's surface; it powers winds, ocean currents, and the water cycle.

   b. Students know solar energy reaches Earth through radiation, mostly in the form of visible light.

   c. Students know heat from Earth's interior reaches the surface primarily through convection.

   d. Students know convection currents distribute heat in the atmosphere and oceans.

   e. Students know differences in pressure, heat, air movement, and humidity result in changes of weather.

5. **Organisms in ecosystems exchange energy and nutrients among themselves and with the environment. As a basis for understanding this concept:**

   a. Students know energy entering ecosystems as sunlight is transferred by producers into chemical energy through photosynthesis and then from organism to organism through food webs.

   b. Students know matter is transferred over time from one organism to others in the food web and between organisms and the physical environment.

   c. Students know populations of organisms can be categorized by the functions they serve in an ecosystem.

   d. Students know different kinds of organisms may play similar ecological roles in similar biomes.

   e. Students know the number and types of organisms an ecosystem can support depends on the resources available and on abiotic factors, such as quantities of light and water, a range of temperatures, and soil composition.

6. **Sources of energy and materials differ in amounts, distribution, usefulness, and the time required for their formation. As a basis for understanding this concept:**

   a. Students know the utility of energy sources is determined by factors that are involved in converting these sources to useful forms and the consequences of the conversion process.

   b. Students know different natural energy and material resources, including air, soil, rocks, minerals, petroleum, fresh water, wildlife, and forests, and know how to classify them as renewable or nonrenewable.

   c. Students know the natural origin of the materials used to make common objects.

7. **Scientific progress is made by asking meaningful questions and conducting careful investigations. As a basis for understanding this concept and addressing the content in the other three strands, students should develop their own questions and perform investigations. Students will:**

   a. Develop a hypothesis.

   b. Select and use appropriate tools and technology (including calculators, computers, balances, spring scales, microscopes, and binoculars) to perform tests, collect data, and display data.

   c. Construct appropriate graphs from data and develop qualitative statements about the relationships between variables.

d. Communicate the steps and results from an investigation in written reports and oral presentations.

e. Recognize whether evidence is consistent with a proposed explanation.

f. Read a topographic map and a geologic map for evidence provided on the maps and construct and interpret a simple scale map.

g. Interpret events by sequence and time from natural phenomena (e.g., the relative ages of rocks and intrusions).

h. Identify changes in natural phenomena over time without manipulating the phenomena (e.g., a tree limb, a grove of trees, a stream, a hillslope).

Mapping Earth's Surface

lesson ❶ Reading Maps

 Grade Six Science Content Standard. 7.f. Students will read a topographic map and a geologic map for evidence provided on the maps and construct and interpret a simple scale map.

● Before You Read

Think about the house or apartment you live in. On the lines below, describe how you would tell someone the best way to drive from school to your house. Then read this lesson to learn more about maps.

_____

_____

_____

MAIN Idea

Maps show large areas of Earth in a size that is easy to read and study.

What You'll Learn
- what latitude and longitude are
- how to use latitude and longitude to determine a location on Earth
- how map scales are used

● Read to Learn

Understanding Maps

People have used maps for hundreds of years. Many types of maps exist. Maps show where things are on Earth or where they are in relation to each other. For instance, some maps might show the locations of streets or landmarks located in a particular city. Some types of maps show the location of parts of Earth's interior. Other maps may show the flow of ocean currents or the positions of the worlds' weather activity.

A map shows the location of things at a given time. Towns, street names, and weather systems can all change. Comparing maps of the same area that have been drawn over a number of years helps you notice the changes that have happened to that area over time.

What are some uses of maps?

Because Earth is so large, maps also help humans determine where they are located on the planet. People use maps to describe their exact position on Earth. For example, when a ship travels across the ocean, the captain uses maps to plot the course. ☑

Study Coach

Identify Main Ideas As you read, underline the main idea in each paragraph.

✔ Reading Check

1. State two uses of maps.

_____

_____

_____

A Sketch and Describe

Make a two-tab Foldable. Label the tabs, as illustrated, then describe and sketch examples of longitude and latitude lines on the front tabs. Describe the importance of each under the tabs.

Picture This

2. Locate Highlight or trace over the equator in red.

What are latitude and longitude?

Mapmakers view Earth as a sphere covered with an imaginary grid of lines that circle the globe. Two sets of lines called latitude and longitude make up this imaginary grid. You can use these lines to find any location on Earth.

Lines of **longitude** run vertically from the north pole to the south pole. Longitude is the distance in degrees east or west of the prime meridian. The prime meridian runs from the north pole to the south pole and runs through Greenwich (GREN ihtch), England. The prime meridian, shown below, represents zero degrees longitude.

On the other side of Earth, directly opposite the prime meridian is a line called the 180° meridian. If you travel 180° degrees east or 180° degrees west of the prime meridian, you reach the 180° meridian.

Each of these lines is a semicircle. Together, the prime meridian and the 180° meridian form a complete circle that divides Earth into two imaginary halves—the eastern hemisphere and the western hemisphere.

Latitude is the distance in degrees north or south of the equator. At Earth's center, a line of latitude called the equator divides Earth into northern and southern hemispheres. Latitude lines form complete circles. The equator forms the largest circle. Notice in the figure below that the circles get smaller as you get closer to the north pole and south pole.

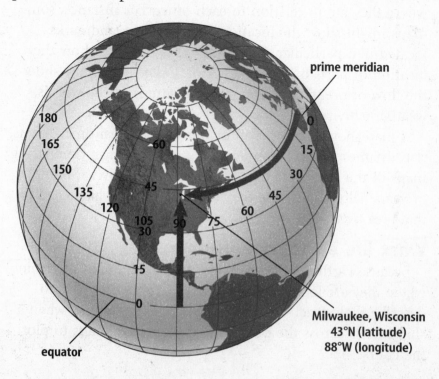

Milwaukee, Wisconsin
43°N (latitude)
88°W (longitude)

What are degrees?

Lines of latitude and longitude are <u>labeled</u> in units called degrees (°). A sphere has 360°. For that reason, each set of lines of latitude and longitude make up 360°. A hemisphere, which represents half the Earth, is divided into 180°.

The north pole is located at 90°N (north) latitude. The south pole is located at 90°S (south) latitude. Globes usually label lines of latitude and lines of longitude every 10°.

What are minutes and seconds?

Latitude and longitude give people a general idea of locations on Earth. More precise locations between lines of latitude or lines of longitude are measured in minutes and seconds. Each line or degree of latitude and longitude is divided into 60 units, called minutes ('). Each minute is also divided into 60 units, called seconds ("). The degrees, minutes, and seconds of a line of latitude or line of longitude can be used to identify an exact location on a map. ☑

How do mapmakers plot locations?

Latitude lines and longitude lines cross and form a huge imaginary grid over Earth's surface. Each intersection of a latitude line and longitude line occurs at an exact point on Earth's surface. The combination of the latitude number and the longitude number is called a coordinate.

Latitude is always listed first when describing a location. For example, the coordinate for locating Sacramento, CA is 38°N (north) latitude by 121°W (west) longitude. You can find California's state capitol building at exactly 38°34′33″N latitude by 121°29′29″W longitude.

How are map view and profile view different?

Most maps are drawn in **map view**, which shows Earth's surface from above, as though you were looking down on Earth's surface. A map view is horizontal, or parallel, with Earth's surface. Map view can also be called plan view.

Sometimes maps are drawn in **profile view**, meaning they are cross sections showing a vertical section of the ground. Think about a profile view like looking at the side of a house, rather than viewing it from above.

Map views are used to describe topographic and geologic maps. Profile views can be used to study models of the inner structures of volcanoes. ☑

Academic Vocabulary
label (LAY buhl) (verb) to describe or identify with a word or phrase

☑ **Reading Check**

3. **Determine** How do you measure latitude and longitude?

_____

_____

_____

💡 **Think it Over**

4. **Classify** What kind of map view would be used to shows roads? (Circle your answer.)
   a. a map view
   b. a profile view

Map Scales and Legends

Maps have two features to help you read and understand the map. One feature is a map legend. The other feature is a map scale.

What do map legends show?

Maps use certain symbols to stand for particular features on Earth's surface. If you pick up any map you will notice a block or box of symbols located somewhere on the map. This is called a map legend. A **map legend** lists all the symbols used on a map so readers can understand and identify what each symbol means. ☑

What are map scales?

When mapmakers draw a map, they need to decide how big or small to make the map. They need to decide on the map's scale.

The map's scale tells you the relationship between a distance on the map and the actual distance on the ground. You may see the phrase, "1 centimeter is equal to 1 kilometer." The scale can also be written as a ratio, so if one centimeter on a map represents one kilometer on the ground, the ratio will be 1:100,000. If you drew a map of your school at a ratio of 1:1, your map would be as large as your school.

The Usefulness of Maps

Maps help people find places on Earth. Latitude and longitude help you to discover the exact location of a place, especially as you focus your search on degrees, minutes, and seconds.

Map legends offer you the key to interpret the map you are using. Map legends explain all the symbols used on the map. Maps scales tell you the relationship between distances on the map and the actual surface distances on Earth. Scales help you determine how far you are going and how long it will take you to get there. ☑

Map views and profile views can help you actually visualize your surroundings and help you find your way around. The type of view you need depends on your purpose for using the map.

You can look at older maps and current maps of your neighborhood and notice changes that have happened over the years. As Earth changes and populations grow, maps will continue to help people plan for the future.

✔ Reading Check

5. Conclude Where would a reader look to find out about a symbol on a map?

_____

✔ Reading Check

6. Identify Which part of a map is the key to the map's distances? (Circle your answer.)
a. map scale
b. map legend

Science Online   ca6.msscience.com

Mapping Earth's Surface

lesson 2 Topographic and Geologic Maps

Grade Six Science Content Standard. 7.f. Students will read a topographic map and a geologic map for evidence provided on the maps and construct and interpret a simple scale map.

● Before You Read

Imagine you have found a great treasure, but you need to hide it for now. You decide to bury the treasure and make a map of its location. What needs to be on the map so you can locate your treasure again? On the lines below, list the important things you want to remember and record on your map. Then read the lesson to find out more about surface maps.

_____

_____

_____

● Read to Learn

Topographic Maps

Topography is the study of the landscape. A topographer is interested in the shape of the surface of the land, including elevation and the positions of its features. **Topographic** (tah puh GRA fihk) **maps** show the shape of Earth's surface. They show major features in nature, such as mountains, lakes, rivers, and coastlines. Topographic maps also show features created by people, such as roads, cities, towns, and buildings. City planners, engineers, the military, scientists, and recreation specialists use topographic maps in their work.

What are contour lines?

Contour lines are the lines drawn on topographic maps that join points of equal elevation. On topographic maps, contour lines tell you the elevation of land as the distance of that land above and below sea level. Contour lines let people measure things like a mountain's height, or an ocean's depth. Each contour line represents a specific elevation on a map. This means the lines never cross.

MAIN‹Idea
Specialty maps show specific features of an area.

What You'll Learn
■ how people use topographic maps
■ how to understand geologic maps
■ the differences between topographic and geologic maps

▸ Mark the Text

Answer Questions As you read the lesson, underline the answer to each question heading.

FOLDABLES

Ⓑ **Describe** Make a Venn diagram Foldable. Record information about topographic maps and geologic maps under the appropriate tabs. Use what you learn to describe what the maps have in common under the center tab.

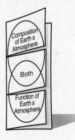

Academic Vocabulary

select (suh LEKT) (verb) to choose

✔ **Reading Check**

1. Identify What is the contour interval on a topographic map?

_____

_____

_____

✔ **Reading Check**

2. Identify What do topographic mapmakers represent with the color red?

_____

_____

What is a contour interval?

The difference in elevation between contour lines that are next to each other is called the contour interval. Mapmakers select a contour interval carefully. They want to emphasize the general shape of a terrain and not overcrowd the map with too many lines. Some contour lines look darker or wider than others. These are called index contours. The index contour is the line that usually has the elevation written on it. ✔

What is a topographic profile?

Sometimes you can get a better idea of the topography of Earth's surface from a cross section or profile view. To read a map in profile view, imagine yourself standing on the ground and looking towards the horizon at the hills and valleys around you.

How are topographic map symbols read?

To understand a topographic map, a person must know how to read the symbols in a topographic legend. Different colors depict specific land features, such as lakes, streams, roads, houses, and even contour lines. The United States Geological Survey (USGS) maps use the color brown to identify contour lines. Lakes, streams, irrigation ditches, and other water-related features are shown in blue. Vegetation is always green and important roads are red. Black is used to identify other roads, bridges, railroads, boundaries, and trails. Some USGS topographical maps use the color purple to identify updated information made to an area since the map was first printed. ✔

The USGS also uses a variety of symbols to identify important information on their maps. An individual house may be a simple, small black square. Larger buildings such as the Rose Bowl in Pasadena are shown with an outline of their actual shape. A tint of color, such as pink, is used to identify large cities or areas of high population. Contour lines that form a V-shape often tell you the locations of streams and erosion channels. Closely spaced contour lines tell you steep slopes are present and widely spaced contour lines show areas where the land is fairly flat.

Knowing these symbols and understanding what they represent are the tools you need to read a topographic map. For example, if you are a hiker, you might use a topographic map to locate the nearest campground or river.

Geologic Maps

Geology is the study of the life of Earth, especially as recorded in rocks. A geologist is interested in identifying areas with landslides, groundwater in wells, types of soils, earthquakes, and valuable minerals. **Geologic maps** help map readers understand the geology of a particular area.

What do colors represent on geologic maps?

Geologic maps are filled with color. The different colors make geologic maps easier to read and understand. Each color represents a geologic formation or rock unit. A **geologic formation** is a three-dimensional body, or volume, of a certain kind of rock of a given age range. For example, sandstone of one age might be represented by a light yellow, while sandstone from another age range might be colored bright orange. ☑

The colors on a geologic map represent the rock units that are nearest to the surface. They are not intended to show actual rock colors. The colors are only used to separate the rocks into different formations. A map key matches the colors in a geologic map to the types of rock they represent.

What are contacts?

Rocks form over, under, or beside other rocks. The place where two rock formations exist next to each other is called a **contact**. Geologists identify two main types of contacts: depositional contacts and fault contacts.

Think about a cliff made of many different layers of rock. The place where these layers touch is called a deposition contact and is represented on a geologic map by a thin black line.

Sometimes, layers of rock break apart or move. When rock formations that are next to one another have been moved the contact is called a fault contact. Fault contacts are represented by a thick black line on geologic maps. Single rock units sometimes break. This means that the same rock unit can lie on both sides of a fault contact. ☑

How are rock units named?

Geologists collect and study many rock samples from Earth's surface. They use the data they collect to create geologic maps. Rock units are often named according to where they are best viewed or studied, like the Briones sandstone that was first found in Briones Valley in California.

☑ **Reading Check**

3. Explain How do you identify a type of rock of a given age range on a geologic map?

_____

_____

☑ **Reading Check**

4. Define What does the thick black line on a geologic map tell you about a rock formation?

_____

_____

How do geologists draw maps of what is underground?

Sometimes, scientists can study layers of rock on Earth's surface, like cliffs or outcrops where rocks are exposed. Often geologists have to drill deep into Earth's surface to get samples of rock. The samples are shaped like long tubes and are called cores. Core samples are studied and the data that is found is recorded for scientific research.

What does a geologic cross section show?

A geologic cross section shows how rocks are stacked under Earth's surface. Rocks form in broad, flat layers called beds. These beds stack up just like the layers of a cake. Sometimes layers remain flat. Other times they tilt or bend. A cross section, such as the one shown below, can help geologists understand how these rock layers exist under Earth's surface.

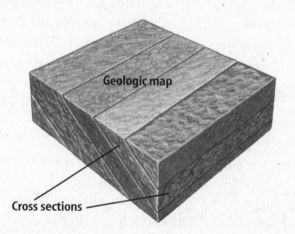

Geologic map

Cross sections

Geology and Maps

Topographic maps and geologic maps are necessary tools for geologists and helpful tools for those who wish to know more about a specific <u>area</u>. Topographic maps show Earth's sloping terrain and wide flat areas. They identify major features on the land's surface, such as roads and buildings and mountains, rivers, and lakes. Geologic maps show what types of surface rock exists in an area, and rock formation and layering.

Together these maps provide a complete picture of the geology of a particular area. Learning to read geologic and topographic maps provides you with the ability to better understand the area you live in or the area you are visiting.

Think it Over

💡 **Think it Over**

5. **Conclude** Why would a cliff be easier to study than rock samples taken by drilling into the earth?

_____

_____

_____

Picture This

6. **Identify** Draw arrows to show the direction of the layers in this cross section.

Academic Vocabulary
area (ER ee uh) (noun) a geographic region

Earth's Structure

lesson ❶ Landforms

 Grade Six Science Content Standard. 1.f. Students know how to explain major features of California geology (including mountains, faults, volcanoes) in terms of plate tectonics. **Also covers:** 1.e, 2.a, 2.b, 2.c.

● Before You Read

Think of a time you went on a trip. Did you see mountains, hills, lakes, rivers, or grassy fields? Describe the landscape on the lines below. Read the lesson to find out more about how landforms are formed.

_____

_____

MAIN ⟨Idea

Forces inside and outside Earth produce Earth's diverse landforms.

What You'll Learn

■ how landforms are created
■ landform varieties in California

● Read to Learn

How do landscapes form?

The surface, or topography, of Earth is always changing. The topography is changed both by forces deep inside Earth and by forces on the surface of Earth. The uneven heating of the inside of Earth pushes up mountains, such as the one below. The heating of Earth's surface by the Sun causes weather, which changes the landscape through erosion. Weather conditions wear down Earth's surfaces, especially in higher areas.

◄ Mark the Text

Identify Main Ideas As you read, underline or highlight one or two main ideas in each paragraph.

Picture This
1. **Circle** the weather on the mountaintop.

Forest

Ocean

Desert

Landforms

Features molded by processes on Earth's surface are called **landforms**. Landforms can cover large regions or can be smaller, local areas. Three main types of landforms are mountains, plateaus, and plains. Mountains and plateaus are areas with high elevations. Plains are low, flat areas. ☑

What landforms are made by uplift?

Uplift is any process that moves the surface of Earth to a higher elevation. Thermal energy inside Earth produces uplift. Thermal energy inside Earth moves toward the surface causing matter inside Earth to also move upward.

Sometimes thermal energy inside Earth melts rocks. A mountain called a volcano can form if this melted rock reaches the surface. Other times the heat does not melt rock. Instead it pushes solid rock upward. Uplift can form new mountains or make existing mountains taller.

When uplift leaves a large flat area, we call the landform a plateau. When uplift leaves a steep landscape, it has formed a mountain.

What landforms are shaped by surface processes?

While thermal energy inside Earth pushes up the land, surface processes wear it down. Energy from the Sun drives these surface processes. Water, wind, and ice break up rocks that form mountains. Mountains get smaller because gravity makes the broken pieces of rock fall and roll downhill.

Erosion The wearing away of soil and rock is called **erosion**. Water in streams and rivers erodes rock fragments and carries them downhill. Rivers flow into lakes and oceans. They cut valleys and steep-sided canyons as the water rushes through the land.

Rivers slow as they get to flatter land. They deposit some of their load of rock and soil. The pieces of rock and soil build new landforms. Beaches are landforms made from rock pieces that were washed by waves from the ocean. Ocean waves can move the rock fragments along the coastline.

✔ **Reading Check**

2. Define What is a landform?

_____

_____

_____

FOLDABLES

Ⓐ Take Notes Make a Foldables table. Use the table to take notes on what you learn about uplift and erosion and the landforms they produce.

	Resulting Landforms	California Examples
Uplift		
Erosion		

California Landforms

California has many types of landforms. Some landforms are so special that they are preserved in state or national parks, such as Yosemite Valley. Glaciers cut a U-shape across Yosemite National Park about one million years ago. When rivers erode Earth's surface, they usually cut V-shaped valleys.

How was Lassen Peak formed?

Lassen Peak is a volcano in California's Lassen Volcanic National Park. It is also part of the Cascade Mountain Range. In 1915 a series of violent eruptions melted rock, gas, and ash on Lassen Peak. The eruption blasted out a new crater in the volcano and created a dramatic change in the surrounding landscape.

Academic Vocabulary
range (RAYNJ) (noun) a series of things in a line; a series of mountains

How were other California mountains formed?

The Sierra Nevada Mountains and the Coastal Ranges are major landforms in California. These mountains were formed by the force of plate tectonics. Solid rock was pushed up to form high peaks. The California mountains form a long and narrow range, sometimes called mountain belts. ☑

Mount Shasta is a volcano. It is different from most mountains in California that were made by uplift. Mount Shasta's cone-shape was formed when melted rock poured out from the center onto the land surface.

California's mountains continue to grow higher. Mountains usually grow so slowly you cannot see the uplift. At times, a volcano explodes or an earthquake causes sudden uplift which can be seen.

✔ Reading Check

3. **Identify** What is a mountain belt?

_____

_____

_____

What makes California's valleys?

Flat valleys are next to the California mountain ranges. The mountains keep pushing higher and erosion keeps wearing the tops down. Wind, water, ice, and gravity work to break off pieces of the mountain. The force of water moves the loose rocks and soil from the mountains down to the valleys. This loose material makes soil rich in nutrients. California's valleys contain some of the best land in the United States for growing plants.

California has many deep, narrow valleys. Rivers carve these valleys as they flow from the mountains to the Pacific Ocean. The water carries small rocks and sand from the Sierra Nevada Mountains through the Central Valley and on to the Pacific coast.

💡 **Think it Over**

4. **Describe** three landforms that are part of California's landscape.

_____

_____

How are beaches formed?

Beaches are <u>temporary</u> features that require constant addition of rocks and sand. The beaches in California are formed from the rocks and sand that wash down from the mountains. The rock, sand, and mud that are moved by rivers are called sediment.

Ocean currents move parallel to the coastline. The currents constantly wash away the sand. If rivers did not keep adding sediment, the beaches would disappear.

Changing Landforms

Landforms may seem permanent but they are changing all the time. Thermal energy from the Sun and from Earth's interior shapes our surroundings and landscapes. Earth's interior energy moves to the surface and creates uplift to form the land into mountains and plateaus. At the same time, Sun's energy creates the weather which wears down the uplifted landforms. Water, wind, ice, and gravity erode the mountains and other landforms.

Sometimes the changes in Earth's surface can be sudden and dramatic. For example, volcanoes erupt and cause sudden changes to landforms. Most changes in Earth's surface happen slowly and steadily. These changes are constantly shaping Earth's landforms.

Academic Vocabulary

temporary (TEM pur rer ee)
(adj.) lasting for a limited time

Think it Over

5. Predict How could a severe drought affect the California coastline?

_____

_____

_____

Earth's Structure

lesson ❷ Minerals and Rocks

 Grade Six Science Content Standard. 6.c. Students know the natural origin of the materials used to make common objects. **Also covers:** 6.b.

● Before You Read

Think of a time you were walking along and found some rocks. Were some of the rocks you found shiny? Were some of them rough or crumbly? On the lines below, describe the rocks you found. Then read about how looking at a mineral's properties can help you identify it.

_____

_____

MAIN ❬Idea❭

The solid Earth is made of minerals and rocks.

What You'll Learn
- the importance of minerals
- how to identify rocks and minerals
- how the rock cycle recycles Earth's materials

● Read to Learn

What is Earth made of?

The solid part of Earth is made up of minerals and rocks. People use them to build homes and roads. Rocks and minerals break down to form the soil in which farmers grow food. Some rocks and minerals, such as diamonds and rubies, can be made into beautiful jewelry. You see rocks and minerals all around you every day.

What is a mineral?

The word *mineral* has several meanings. Most minerals form from substances that were never living. But there are minerals made by living things. For example, aragonite is a mineral found in pearls, which are made by oysters. Scientists have identified about 3,800 different minerals. Most of these minerals are rare. There are only about 30 common minerals.

In Earth science, the word *mineral* has a specific definition. A **mineral** is a naturally occurring, generally inorganic solid that has a crystal structure and a definite chemical composition. The characteristics of minerals are summarized in the table on the next page.

◀ **Mark the Text**

Identify Unknown Words As you read, underline any term or word that you don't understand. Look up the meaning of these words.

FOLDABLES™

B **Define** Use two sheets of notebook paper to make a layered Foldable. As you read the lesson, define and record what you learn about minerals, rock groups, and the rock cycle under the tabs.

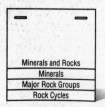

Picture This

1. Interpret According to the chart, when is water a mineral?

_____

_____

Mineral Characteristics		
Characteristic	**Explanation**	**Example**
Naturally Occurring	• The substance must be found in the natural world.	• Diamonds mined from Earth are minerals; artificial diamonds created in a laboratory are not minerals.
Generally Inorganic	• The substance is not a living plant or animal. • Some minerals are made by living things.	• The mineral aragonite is found in pearls, which are made by oysters. • The mineral apatite is found in bones and teeth.
Solid	• Minerals are solid and keep their own shape. • Gases and liquids are not minerals.	• Water in a glacier is a mineral; water from a melting glacier is not a mineral.
Crystal Structure	• Each mineral is made of specific atoms arranged in orderly, repeating patterns.	• Smooth faces on a crystal represent a well-organized internal structure of atoms.
Definite Composition	• A mineral is made of specific elements. • The elements of a mineral are in definite proportions, which can be expressed in a chemical formula.	• The chemical formula for quartz is SiO_2, meaning it has two oxygen atoms for every silicon atom.

Physical Properties of Minerals

You can tell one mineral from another by its physical properties. Physical properties are characteristics that can be observed or measured without changing the identity of the mineral. Each mineral has physical properties that are used to identify or name it. You can learn how to test a mineral for these properties and then use the tests to identify many minerals. Some common physical properties used to test and identify minerals are described next. ☑

2. Describe What is used to identify a mineral?

_____

_____

_____

14 Chapter 2 Earth's Structure

Reading Essentials

What is mineral hardness?

A measure for how easily a mineral is scratched is its hardness. In the early 1800s, Friedrich Mohs, an Austrian scientist, developed a scale that compares the hardness of minerals. Mohs' hardness scale was based on 10 minerals with a hardness rating from 1 to 10. Diamond was the hardest mineral, with a score of 10. The softest mineral, talc, rated a hardness score of 1. The Mohs hardness scale, shown in the table below, is still used today to identify gemstones in jewelry stores.

Mohs Hardness Scale		
Mineral	**Hardness**	**Common Tests**
Talc	1	rubs off on clothing
Gypsum	2	scratched by fingernail
Calcite	3	barely scratched by copper coin
Flourite	4	scratches copper coin deeply
Apatite	5	about same hardness as glass
Feldspar	6	scratches glass
Quartz	7	scratches glass and feldspar
Topaz	8	scratches quartz
Corundum	9	scratches most minerals
Diamond	10	scratches all common materials

Why can color not always be used to identify minerals?

A mineral's color will sometimes help you identify it. The mineral malachite always has a unique green color, because it contains copper. Most minerals, however, come in several colors. For instance, diamonds can be white, blue, yellow, or pink. Many other gemstones are these colors, too.

What are streaks?

Streak is the color of powder from a mineral. Some minerals that vary in color have distinct streak colors. A particular mineral could be silver, black, brown, or red in color. Yet when each of these colors of the mineral is rubbed across an unglazed porcelain tile, you will discover the same colored streak. In other words, the streak of a mineral will be the same color no matter what the color of the solid mineral.

Picture This

3. Name What mineral scratches glass?

_____

Academic Vocabulary
instance (IHN stunts) (noun) an example; a case or illustration

What is luster?

Luster is the way a mineral's surface reflects light. Geologists study minerals and rocks. They use several different words to describe mineral luster. Some of the terms for luster are *greasy*, *silky*, and *earthy*. ☑

What are the shapes of minerals?

Every mineral has a unique crystal shape. A crystal that forms on Earth's surface will be small, because the erupting lava cools rapidly. Crystals are large and perfect when they form underground where Earth's heat is maintained and the magma source cools slowly. Each crystal has a particular shape, which sometimes is referred to as its crystal habit.

What are cleavage and fracture?

Cleavage and fracture describe the way a mineral breaks. If a mineral breaks along smooth, flat surfaces, it has cleavage as shown in the figure of calcite below. When a mineral has cleavage it feels smooth and glassy.

A mineral can break so that flat surfaces are seen in more than one direction, shown in the figure of quartz below. A mineral that breaks along rough or irregular surfaces displays fracture. When a mineral fractures it feels rough to your hand.

Quartz

Calcite

What is density?

The amount of matter an object has per unit of volume is called **density**. Some minerals are denser than others. For example, if you pick up a piece of galena and a piece of quartz the same size, you can feel that the galena is much heavier. This is because galena is denser than quartz.

Minerals, such as metals, with atoms that are packed closely together also tend to have higher densities. Density can be used to help identify minerals. ☑

Reading Check

4. Name What do we call the reflection of light from a mineral's surface?

_____

Picture This
5. Determine Draw a circle around the mineral with fracture.

Reading Check

6. Explain Why are some minerals heavier than others of the same size?

_____

_____

What are other properties of minerals?

Some minerals have properties that make them easy to identify. For example, magnetite is magnetic. Calcite shows an interesting property that occurs when light interacts with it. If you look at an object through a clear calcite crystal, you can see two <u>images</u> of the object. Graphite can be used to mark paper. Copper is a good conductor of electricity. Every mineral has properties that can be observed to help identify it.

Mineral Uses

Rich deposits of valuable minerals are called ores. The metals you use every day can be extracted from these ores. Iron, used to make steel, comes from hematite and magnetite ores. Steel is used to manufacture such items as cars, bridges, and skyscrapers. The table below lists some ores and their uses.

People appreciate some minerals solely for their beauty. These minerals are known as gemstones. Many gemstones have intense colors, a glassy luster, and are high on the Mohs hardness scale. Diamonds, rubies, sapphires, and emeralds are among the most valuable gemstones. These rare gemstones can be cut, polished, and set into jewelry.

Common Uses of Minerals and Ores		
Ores	**Minerals**	**Mineral Uses:**
Chalcopyrite Malachite	copper	• metal in wires to conduct electricity • copper roofing and gutters • jewelry, housewares
Hematite Magnetite	iron	• manufacture of cars • building bridges and skyscrapers • everyday home products
Galena	lead	• automobile battery parts • radiation shields
Gold ore Silver ore	gold and silver	• computer circuits, electronics • air bags in cars • jewelry settings and chains • decorative furnishings
Bauxite ore	aluminum	• alloy rims and other parts for cars • bikes, airplanes • Soft drink cans and food containers

Picture This

7. **Name** a product that is made from lead.

_____

_____

_____

Rocks

A **rock** is a natural, solid mixture of particles. These particles are mainly made of individual mineral crystals, broken bits of minerals, or fragments of rocks. Rocks may also contain pieces of dead animals, shells, bones, and volcanic glass. The pieces that make up a rock are called grains.

Mountains, valleys, and even the seafloor under the oceans are made of rocks. Rocks are classified into groups based on how they form. There are three major groups of rocks: igneous rocks, metamorphic rocks, and sedimentary rocks.

What are igneous rocks?

Igneous rocks are formed from molten, or liquid, rock material called **magma**. As the hot magma from deep underground moves closer to Earth's surface the magma begins to cool. Tiny crystals of minerals form as the magma becomes a hard solid. These individual crystals become the grains in an igneous rock. ✔

Magma that comes to the surface is called **lava**. Volcanic glass forms when lava cools so rapidly that the atoms cannot organize into crystal structures.

Granite is an igneous rock that cools slowly and has larger grains. Basalt is an igneous rock that cools more rapidly and has very tiny grains. The grain size and the way grains fit together in a rock are called texture.

How do metamorphic rocks form?

Metamorphic rocks form when solid rocks are squeezed, heated, or come into contact with fluids. Metamorphic rocks remain solid as they change. The original rock that is changed is called the parent rock. Heat, pressure, or hot fluids applied to the parent rock causes new mineral grains to grow. The new grains have a different texture and mineral makeup than the parent rock. When a parent rock is put under pressure from one direction, the grains can form layers that look like stripes. This striped pattern is called foliation.

How do sedimentary rocks form?

Rocks are broken down by physical and chemical changes on the surface of Earth. **Sediment** is rock that is broken down into smaller pieces or dissolved in water. Sediment eventually settles in low-lying areas. Sediment is changed to sedimentary rock as grains are forced together by the weight of new layers of sediment. ✔

✔ Reading Check

8. Identify the material from which igneous rock is formed.

_____

_____

✔ Reading Check

9. Define What is sediment?

_____

_____

_____

_____

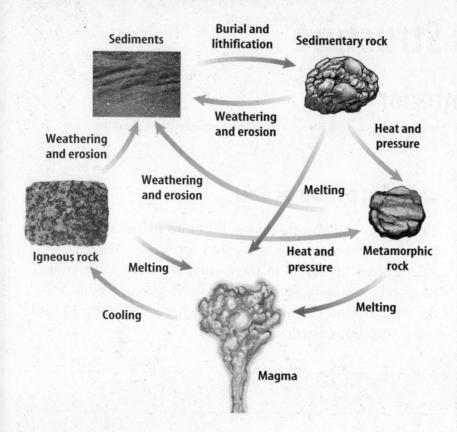

Sediments

Burial and lithification

Sedimentary rock

Weathering and erosion

Heat and pressure

Weathering and erosion

Weathering and erosion

Melting

Melting

Metamorphic rock

Igneous rock

Melting

Heat and pressure

Cooling

Melting

Magma

Picture This
10. **Identify** Use colored pencils to add information to the rock cycle. Color arrows that involve heat red, weathering and erosion green, cooling blue, and compaction brown.

What is the rock cycle?

The <u>rock cycle</u> is a series of processes that changes one type of rock into another rock. Forces on Earth and inside the planet provide the energy for the rock cycle.

The three major rock groups—igneous, metamorphic, and sedimentary—are related through the rock cycle, shown above. Different Earth materials—magma, sediment, and rocks—are connected by arrows that show the many ways rocks change. It takes thousands to millions of years for rocks to move through the rock cycle.

Earth Materials

The solid part of Earth is made up of minerals and rocks. The properties of minerals include hardness, luster, streak, color, crystal habit, cleavage, or fracture. Scientists use these properties to help identify a mineral.

Igneous rocks are made from melted rock that moves from Earth's interior to the surface. Metamorphic rocks form when rocks are exposed to pressure or heat and change without melting. Sedimentary rocks form when sediment is pressed together. The rock cycle describes the ways in which rocks change from one type to another type.

Think it Over

11. **Compare** How is recycling plastic similar to the rock cycle?

_____

_____

_____

_____

Earth's Structure

lesson 3 Earth's Interior

Grade Six Science Content Standard. 1.b. Students know Earth is composed of several layers: a cold, brittle lithosphere; a hot, convecting mantle; and a dense, metallic core. **Also covers:** 4.c.

MAIN Idea

Earth's interior has a layered structure.

What You'll Learn

- the major layers of Earth
- the role that convection plays inside Earth

Study Coach

Outline Create an outline of this lesson as you read. Use the headings in the lesson as the main points in the outline.

Reading Check

1. Describe How do seismic waves help us learn about the inside of Earth?

_____

_____

_____

_____

● Before You Read

Take a trip and you might see hills and mountains that have been cut away to make room for a road. Have you ever noticed the layers in these cuts? Describe a cut-away mountain you have seen. Where was it located? What was the land around the mountain like? Read the lesson to find out more about Earth's interior.

_____

_____

● Read to Learn

Layers

Earth's interior is made up of three layers. Each layer is different due to differences in temperature and pressure. The deeper you go into Earth's interior, the greater the temperature and pressure. Each layer is also made up of different materials.

What do seismic waves tell us about the Earth?

Scientists have removed samples of Earth's interior to a depth of about 12 km. Volcanoes provide rock samples from as deep as 200 km into Earth's interior.

Scientists need other methods to determine what the deepest parts of Earth's interior are like. Earthquakes produce seismic waves that pass through the planet. The speed and direction of seismic waves change depending on the type of material they pass through. The waves bounce off or bend as they approach a new layer. Scientists learned about the layers and insides of Earth by studying the paths of seismic waves. As the scientists study the paths of the seismic waves, they can discover new details about the makeup of Earth's interior. ✓

What is the crust of Earth?

The **crust** of Earth is the thin, rocky, outer layer. Crust under the oceans is made of the igneous rock called basalt. Below the basalt is another igneous rock called gabbro. Gabbro (GAH broh) has the same composition as basalt, yet has larger grains because it cooled more slowly.

Most of the continental crust is made of igneous rocks made of low-density minerals, such as granite. This makes the average continental crust less dense than oceanic crust. The crust's igneous rocks are usually covered with a thin layer of sedimentary rocks. Rocks that make up the crust are rigid and break easily.

What is the mantle of Earth?

The **mantle** is the thick middle layer of rock below the crust. The rock in the upper part of the mantle is called peridotite (puh RIH duh tite). Minerals in mantle rocks have tightly packed crystal structures, making mantle rocks denser than rocks in Earth's crust.

The mantle is made of two layers caused by increasing temperature and pressure. Rocks in the upper mantle are brittle. Between 100 km and 250 km deep it is so hot that tiny bits of the rock melt. This partly melted rock material allows the rock to flow. Scientists sometimes call this flowing rock *plastic*. This flowing, but still mostly solid layer of the mantle is called the **asthenosphere**. If it were possible for you to visit the mantle, you would not see the flow. It moves at rates of only a few centimeters per year. ☑

Below the asthenosphere, the rock is solid, even though it is hotter than the rock material in the asthenosphere. How can this happen? The pressures deep within Earth are so great that they squeeze hot rock material into a solid state.

What is the lithosphere?

The rock in the crust and mantle has different compositions. But the crust and mantle are both made of solid and rigid rock. Together, the crust and the uppermost mantle form the brittle outer layer of Earth called the **lithosphere**.

What is the core of Earth?

The dense metallic center of Earth is called the **core**. It is made mainly of iron and nickel metals. The core is divided into two layers. The outer core is a layer of molten metal. Higher pressures in the inner core cause the metal to be in the solid state.

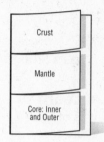
✔ Reading Check

2. Determine What allows rock to flow?

_____

_____

Heat Transfer in Earth

The heat movement in a fluid is by a process called convection. Convection moves heat in the outer core and in the mantle. This movement process is the result of changes in density.

What factors affect Earth's density?

The density of Earth's materials varies. Some minerals and rocks are denser because of their composition. Temperature and pressure can also affect density.

As the temperature of a material is raised, the volume increases. The mass of a material does not change, but it takes up more space, so it is less dense. As pressure is raised, the material is squeezed into a smaller space. This causes the material's density to increase. ☑

The three major layers of Earth are made of different materials with different densities. The core is metallic. The mantle and crust rocks are less dense than material in the core. The rocks in the crust are the least dense of all rocks.

How does convection affect the core and mantle?

Heat energy in Earth's outer core and mantle moves toward the surface mostly by convection. This is important for two major Earth processes. First, convection in the outer core produces Earth's magnetic field. This causes Earth to act a little like a huge bar magnet.

Second, convection in the mantle is important for plate tectonics. Scientists have discovered that even solid rock can flow. In order for this to happen, the rock must be very hot in some places and cooler in other places. The flow takes place extremely slowly. The flow transfers energy and matter from the mantle to Earth's plates. Recent studies show that the plates might control the flow of the mantle below them. There still is much to learn about this movement of material in Earth's interior.

What have you learned?

Now that you've thought about Earth's structure from the surface to the core, you probably realize that Earth is a <u>dynamic</u> planet. Material within Earth continues to move because it is energized by the decay of radioactive elements deep inside the planet. As movement of matter occurs, heat escapes and changes Earth's surface by uplift.

✔ **Reading Check**

3. **Identify** two factors that affect the density of rocks.

_____

_____

Academic Vocabulary
dynamic (di NA mihk) (adj.)
constantly moving or changing; energetic

Science  nline   ca6.msscience.com

Thermal Energy and Heat

lesson ① Forms of Energy

Grade Six Science Content Standard. 3.a. Students know energy can be carried from one place to another by heat flow or by waves, including water, light and sound waves, or by moving objects.

● Before You Read

Think of cookie dough before you put it in the oven to bake. On the lines below, describe how it is different from the dough that has been baked. Then read this lesson to find out how energy creates change.

What You'll Learn
■ about the different forms of energy
■ the difference between kinetic energy and potential energy

● Read to Learn

What is energy?

What changes do you see around you? You feel the wind on your face. You see cars moving. You walk the halls of your school. All the changes around you are caused by energy. **Energy** (EN ur jee) is the ability to cause change and it exists in many forms.

What is kinetic energy?

Think of a baseball traveling through the air. The energy the ball has is called **kinetic** (kuh NEH tihk) **energy**. Kinetic energy is the energy an object has because it is moving. An object that is not moving does not have kinetic energy. The kinetic energy of an object depends on two things. One is the object's mass. The other is the object's speed. The kinetic energy of an object increases if the mass or speed of the object increases.

If an object has a large mass and moves at a slow speed, it can have a lot of kinetic energy. For example, a glacier is a large mass of ice that has a very slow speed. It may move only a few meters a year. Even so, glaciers have enough kinetic energy to change the land as they move over it. Yosemite Valley in California was changed by the kinetic energy of slowly moving glaciers.

Mark the Text

Highlight As you read this lesson, underline or highlight the definition of each vocabulary term.

FOLDABLES

Ⓐ **Take Notes** Make a Foldable table and label as illustrated. Use the table to take notes on what you learn about kinetic and potential energy.

	Define	Forms of Energy	Calculate
Kinetic Energy			
Potential Energy			

Potential Energy—Stored Energy

Would you say that a rock has energy? An object such as a rock can have energy even if it is not moving. Energy is the ability to cause change, so a rock has energy. When the rock falls, it causes a change. Even before the rock was falling, it had energy. The rock has stored energy called **potential** (puh TEN chul) **energy**. There are different forms of <u>potential</u> energy.

What is gravitational potential energy?

A rock hanging above ground has a form of stored energy. It is called gravitational potential energy. This type of potential energy depends on an object's mass and its height above the ground.

Imagine that you accidentally dropped a plastic bottle filled with water on your foot. If the bottle fell from your waist, it might bruise your foot but it probably wouldn't break it. Now imagine the water bottle falling from the top of a roof onto your foot. Would it cause more damage to your foot? The higher an object is above a surface, the greater gravitational potential energy it has. Also, the greater the object's mass, the more gravitational potential energy it has.

What is elastic potential energy?

Have you ever seen a spring that stretches and then pulls back into shape? The energy stored when an object is stretched or squeezed is called **elastic** (ih LAS tik) **potential energy** as shown below. If the spring is squeezed and let go, it likely will return to its original length. If the spring is stretched and let go, it also will return to its original length. Elastic potential energy gives an object the ability to change.

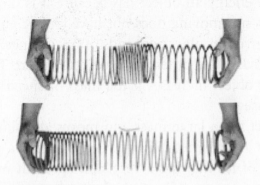

Picture This

1. Predict what will happen to a spring if it is compressed and then let go?

_____

_____

_____

Where is chemical potential energy stored?

When you eat, you take in another type of potential energy. Chemical potential energy is stored in bonds between the atoms that make up matter. Remember that atoms are joined together by chemical bonds to form molecules.

A chemical reaction can release the chemical potential energy stored in chemical bonds. When these bonds are broken, new bonds are formed. You get energy by eating because food contains chemical potential energy. Oil and coal, called fossil fuels, are energy sources because they contain chemical potential energy.

Light Energy and Thermal Energy

When you turn on a lamp, change occurs. Light from the lamp makes it possible for you to see things in the room more clearly. When you turn on a stove to heat a pot of water, change occurs. Heating the pot causes the temperature of the water to increase. These changes are caused by light energy and thermal energy.

What form of energy comes from sunlight?

What causes plants to change? You know that plants need sunlight to grow. Sunlight contains a form of energy called light energy or radiant energy. Light energy is the energy carried by light waves. Light energy spreads out, or radiates, in all directions from its source. Plants change light energy to chemical energy.

What is thermal energy?

If you put your hands around a warm cup of cocoa, your hands will feel warmer. The warmth is caused by thermal energy. **Thermal** (THUR mul) **energy** moves from one place to another because of differences in temperature.

Thermal energy can cause changes. The cup of cocoa has a higher temperature than your hands. The thermal energy, which is sometimes called heat energy, moves from the hot cocoa to your cooler hands. This causes a change to occur. Your hands become warmer and the cocoa becomes cooler. ☑

The Different Forms of Energy

All forms of energy can cause change. A moving object has kinetic energy. Potential energy is energy that is stored. Thermal energy is energy that moves because of differences in temperature.

Academic Vocabulary
occur (oh KUR) (verb) to come into existence; to happen

✔ Reading Check

3. Explain What causes your hands to become warmer from holding a cup of hot cocoa?

_____

_____

Thermal Energy and Heat

lesson ❷ Energy Transfer

Grade Six Science Content Standard. 3.a. Students know energy can be carried from one place to another by heat flow or by waves, including water, light and sound waves, or by moving objects. **Also covers:** 3.b.

MAIN ‹Idea

Moving objects transfer energy from one place to another.

What You'll Learn

■ how waves transfer energy from one place to another

■ how energy can change from one form to another

Study Coach ▶

Record Questions As you read this lesson, write down any questions you have about energy and how it affects your daily life. Discuss these questions with another student or your teacher.

☑ Reading Check

1. Explain Why is no work done when you push on a wall and it does not move?

_____

_____

_____

● Before You Read

On the lines below describe how a ball is moved when you kick it. Then read this lesson to learn more about what happens when your body transfers energy to move a ball.

_____

_____

● Read to Learn

Moving Objects Transfer Energy

How is energy transferred when a pitcher throws a ball? Moving objects transfer energy from one place to another. The moving ball has kinetic energy. The ball transfers this energy to the catcher's mitt.

When is work performed?

When you push or pull on something, you are transferring energy. Scientists define **work** as the transfer of energy that occurs when a push or a pull makes an object move. A push or a pull is also called a force. Work is done only when an object moves in the same direction as the applied force.

If you pull upward on a box, you cause the box to move upward. Your pull is a force that makes the box move, so you have done work. When you increase the height of the box above the ground, the gravitational potential energy of the box increases. By lifting the box, you transfer energy to the box. ☑

Waves Transfer Energy

A **wave** is a disturbance in a material that transfers energy without transferring matter. Like moving objects, all waves transfer energy from one place to another.

What form of energy do water waves transfer?

Think about a time when you may have been floating in the water. When a wave passes by, you move up and down and back and forth, but you have no overall movement in the direction of the wave. Water waves transfer kinetic energy from one place to another, but they do not transfer matter.

How do sound waves transfer energy?

You can see water waves, but sound waves are waves you cannot see. They also transfer energy. Sound waves are caused by the back-and-forth movement, or vibration, of an object.

When a drummer hits a drum, the head of the drum moves up and down many times each second. Each time the drum head moves, it hits nearby air particles and transfers kinetic energy to them. Air molecules then bunch together and spread apart, as shown below. Kinetic energy is passed through the air as sound waves. The sound wave does not carry air particles from the drum to your ear. Like water waves, sound waves transfer energy but not matter.

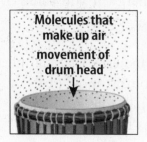
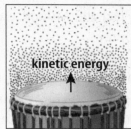
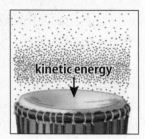

What are electromagnetic waves?

Like sound waves and water waves, light is also a type of wave. However, unlike water waves and sound waves that can only travel in matter, light waves can also travel in empty space. For example, the Sun gives off light waves that travel almost 150 million km to Earth through empty space. Light waves are a type of wave called electromagnetic waves. Electromagnetic waves are waves that can <u>transfer</u> energy through matter or empty space. The energy carried by electromagnetic waves is called radiant energy.

FOLDABLES™

B **Describe** Use one sheet of notebook paper to make a three-tab Foldable. Under each tab describe how water waves, sound waves, and electromagnetic waves transfer energy.

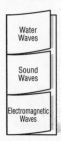

Picture This

2. Sketch What do you think a sound wave would look like if you could see one? Sketch a drumstick hitting a drum and the sound waves, or vibrations, moving out from the drum.

Academic Vocabulary
transfer (TRANS fur) (verb) to move or shift

Energy Conversions

Energy comes in different forms and can be transferred from place to place. Energy can also be converted from one form to another. When energy changes form, it can become more useful.

At what point is energy converted when a ball is tossed up in the air?

When you throw a ball upward, energy changes form. As it moves upward, the ball's kinetic energy changes into potential energy. When the ball reaches its highest point, all its kinetic energy has been converted to potential energy. Then, as the ball falls down, potential energy is converted back into kinetic energy. ☑

What does fuel produce?

Where does energy come from when the wood in a campfire burns? Remember that chemical potential energy is energy stored in the bonds between atoms. Atoms make up matter. Chemical potential energy changes the burning wood into thermal energy and radiant energy. If you stand near the fire, you can feel the radiant energy that is given off by the burning wood.

Wood is an example of fuel. **Fuel** is a material that can be burned to produce energy. When wood burns, most of its chemical energy changes form. Only a small fraction of the wood's chemical potential energy remains in the ashes.

What kind of energy is produced by a car's engine?

As you know, gasoline is burned in a car's engine. As the gasoline burns, most of its stored chemical potential energy changes to thermal energy. The car's engine changes the thermal energy into forces that make parts of the engine move. The car's engine converts thermal energy into the kinetic energy of the moving car.

Some of the thermal energy, however, is not converted into kinetic energy. It does not produce useful work. If you place your hand close to the hood of a recently driven car, you can feel unconverted thermal energy. It is wasted energy because it was not used to move the car. ☑

✔ **Reading Check**

3. Describe When you throw a ball in the air, at what point is kinetic energy converted to potential energy?

_____

_____

_____

What is friction?

What happens when a bike rider applies the bicycle's brakes? The bike shown below slows because its brake pads rub against the wheels. **Friction** (FRIK shun) is the force between surfaces that opposes the motion of an object. It acts between the wheels and the brake pads, causing the bike to slow down. When the brake pads rub against the wheels, most of the kinetic energy changes to thermal energy due to friction. As a result, the bike comes to a stop.

Force due to friction

Force due to friction

What have you learned?

Energy can move from place to place in many ways. Energy is transferred when objects move. Waves transfer energy. Three kinds of waves that transfer energy are water waves, sound waves, and electromagnetic waves.

Energy can also be converted from one form to another form. For example, burning fuels converts chemical potential energy into thermal energy and radiant energy. The chemical potential energy is stored in the bonds between atoms and molecules. Burning the fuel releases the energy in the fuel.

Picture This

5. Identify Trace the arrow that shows the direction of force the brake makes on the wheel. What force is the brake opposing?

_____

_____

Think it Over

6. Compare In terms of energy, how are a person eating food and a car burning gasoline similar?

_____

_____

Thermal Energy and Heat

lesson ❸ Temperature, Thermal Energy, and Heat

Grade Six Science Content Standard. 3.a. Students know energy can be carried from one place to another by heat flow or by waves, including water, light, and sound waves, or by moving objects.

MAIN ❰Idea

Thermal energy flows from areas of higher temperature to areas of lower temperature.

What You'll Learn

- how temperature depends on particle motion
- how to compare different temperature scales

Study Coach ▶

Activate Background Knowledge Skim this lesson and highlight any terms that you may have learned earlier in science, such as mass or volume. If you can't remember the definitions of the terms, ask another student or your teacher.

FOLDABLES™

C **Define** Make a layered Foldable. As you read the lesson, define and record what you learn about temperature, thermal heat, and heat under the tabs of this study guide.

● Before You Read

Have you ever seen an egg cooked in a skillet? On the lines below describe the changes that occur once the egg hits the hot pan and heat energy transfers to the egg. Then read this lesson to find out more about heat.

_____

_____

● Read to Learn

What is temperature?

You know that cooking changes the temperature of food, but what does the word *temperature* really mean? Temperature depends on the movement of the particles that make up matter.

Does matter contain particles in motion?

Look at objects such as desks and chairs that are sitting still. These objects, and all matter, contain particles called atoms and molecules that are always moving. Even though the object may not appear to be moving, the particles that make it up are constantly in motion.

How does temperature depend on particle motion?

Particles can move at many different speeds and in many directions. Some move slowly, while others move fast. What does particle speed have to do with energy and temperature? An object's kinetic energy depends on its speed and mass. If two particles have the same mass, the one that moves faster has more kinetic energy. **Temperature** is a measure of the average kinetic energy of the particles in a material.

What is thermal expansion?

You can't see the movement of <u>individual</u> particles in a material, but sometimes adding thermal energy can change particle motion. If a balloon is sealed so no particles can get in or out and you heat the balloon with a blow drier, the balloon will look like it is being inflated. Why does that happen?

As the temperature of the air in the balloon increases, the particles move faster. The particles run into one another with more energy and take up more space. An increase in the volume of a substance when the temperature increases is called **thermal expansion** (THUR mul • ihk SPAN shun). Most materials expand when their temperature increases. Usually, the greater the increase in temperature, the more the material expands. ☑

Measuring Temperature

Temperature is a measure of the average kinetic energy of the particles in a material. However, these particles are so small that it is impossible to measure their kinetic energies. Instead, a practical way to measure temperature is with a thermometer.

How does a thermometer work?

Thermometers contain a red liquid inside a thin glass tube. The particles in the red liquid will expand as the temperature increases. The red liquid in a thermometer placed in boiling water has a greater volume than when it is placed in ice water. A scale on the glass tube shows the temperature as the red liquid inside the tube rises or falls.

What are temperature scales?

There are three common temperature scales—Fahrenheit, Celsius, and Kelvin. The figure at the top of the next page shows two common temperature scales—Fahrenheit and Celsius. On the Fahrenheit scale, water freezes at 32°F and boils at 212°F. On the Celsius scale, water freezes at 0°C and boils at 100°C. Another scale, the Kelvin scale, registers temperatures as 273 degrees more than temperatures on the Celsius scale. To change from Celsius degrees to Kelvin degrees, add 273 to the Celsius temperature. The United States uses the Fahrenheit scale, but other countries use the Celsius scale. The Celsius and Kelvin scales both are used in science.

✔ **Reading Check**

1. **Explain** Why do most substances increase in volume when their temperature increases?

_____

_____

_____

_____

Applying Math

2. **Determine** If the temperature of water is 0° on the Celsius scale, what would it be on the Kelvin scale?

_____

_____

_____

Picture This

3. Add Labels In the blanks, write the temperature of the freezing points on the Fahrenheit and Celsius scales.

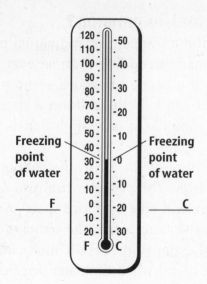

Freezing point of water

Freezing point of water

_____ F   _____ C

Heat

When you put an ice cube in a glass of water, the water becomes colder. The water becomes colder because thermal energy moves from the warmer water to the colder ice cube. The movement of thermal energy from an object at a higher temperature to an object at lower temperatures is called **heat**. ☑

Heat always transfers energy from something at a higher temperature to something at a lower temperature. Suppose a bowl of hot soup sits on a table. After a while, the soup and the air around the soup will have the same temperature. Thermal energy has moved from the hot soup to the surrounding air. Thermal energy keeps moving from warmer objects to cooler objects until both objects are the same temperature.

What have you learned?

Temperature measures the average kinetic energy of particles in a material. As the particles in a material move faster, the temperature of the material increases. The material also expands in <u>volume</u> as the temperature increases.

Heat is the movement of thermal energy due to differences in temperature. Thermal energy always moves from warmer object to cooler objects. Thermal energy stops moving when the objects have the same temperature.

☑ **Reading Check**

4. Define What is heat?

_____

_____

_____

_____

Academic Vocabulary
volume (VAHL yewm) (noun)
amount

Science▶nline   ca6.msscience.com

Thermal Energy and Heat

lesson ❹ Conduction, Convection, and Radiation

Grade Six Science Content Standard. 3.c. Students know heat flows in solids by conduction (which involves no flow of matter) and in fluids by conduction and by convection (which involves flow of matter). **Also covers:** 3.d.

● Before You Read

On the lines below describe why you think builders insulate the outer walls of buildings. Read this lesson to learn more about the purpose of insulation.

_____

_____

_____

● Read to Learn

Conduction

You know that thermal energy moves from one material to another because of differences in temperature. A way that thermal energy moves is called conduction. **Conduction** (kuhn DUK shun) is the transfer of thermal energy by collisions between particles in matter.

How do particle collisions transfer energy?

Collisions transfer energy from particles with more kinetic energy to those with less energy. For example, as heat transfers from hot soup to a cooler spoon, particles in the soup collide with nearby particles in the spoon.

Conduction also occurs within the spoon. Particles in the spoon closest to the soup are the first to gain kinetic energy from the soup. The particles in the spoon then collide with other nearby particles, passing on kinetic energy throughout the spoon.

How does conduction transfer heat in solids?

Conduction transfers thermal energy by collisions of particles. In solids, collisions only occur between particles that are next to each other. Particles in a solid are close together. They move back and forth slightly, but stay in one place. Thermal energy is conducted in solid objects as kinetic energy passes from one particle to another.

MAIN Idea

Thermal energy is transferred by conduction, convection, and radiation.

What You'll Learn

■ how thermal energy is transferred by collisions between particles

■ how thermal energy is transferred by electromagnetic waves

Study Coach

Create Two-Column Notes Learn the answers to the questions that appear as headings in this lesson by writing the question on the left side of your paper and the answer on the right side.

FOLDABLES

❷ **Describe** Use one sheet of notebook paper to make a three-tab Foldable. Under the tabs, describe conduction, convection, and radiation in your own words, and give specific examples of each.

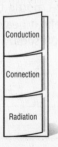

What is a conductor?

A <u>conductor</u> is a material that quickly moves thermal energy. Solids and liquids are better conductors than gases because their particles are closer together. Collisions occur more often in solids and liquids, so kinetic energy can be transferred through material faster. Metals are the best conductors, which is why cooking pans often are made of metal. ☑

What is an insulator?

An insulator is a material that does not transfer heat energy easily. The particles in gases are so spread apart that collisions occur less often. The rate of heat flow in gases is slower than in solids and liquids. Because air is a mixture of gases, thermal energy moves slowly through it. Air is an example of a good insulator.

Convection

Thermal energy is also transferred by convection. <u>Convection</u> (kuhn VEK shun) is the transfer of thermal energy by the movement of matter from one place to another. The particles of a material must be able to move easily from place to place for convection to take place. Solids do not have particles that move easily. A <u>fluid</u> is a material made of particles that can easily change their locations. Convection occurs only in liquids and gases because they are fluids.

What is density?

During convection, parts of a fluid that have a higher temperature move where the temperature of the fluid is lower. Why would a material flow just because the temperature is different? The answer is density (DEN suh tee). Density is the mass contained in a unit volume of a material. When you pick up a full 2-L bottle of soda, it is heavier than picking up a 2-L bottle that is full of air. Both bottles have the same volume. Because the density of soda is greater than air, the bottle of soda is heavier.

Does density depend on temperature?

Different materials have different densities. Sometimes samples of the same material can have different densities. If the temperature of a material increases, it will cause a material to expand. Remember that this is called thermal expansion. The mass of the material doesn't change, but if the volume increases, then its density will decrease.

1. **Explain** Why are solids and liquids better conductors than gases?

_____

_____

_____

💡 **Think it Over**

2. **Draw Conclusions** What material could you put in a 2-L bottle that would make it heavier and denser than soda?

_____

_____

_____

What does density have to do with floating?

Why do some objects sink in water and others float? An object will sink if its density is greater than the fluid surrounding it. An object will float if its density is less than that of the fluid surrounding it. A stone will sink in water because its density is greater than the density of water. A stick will float because its density is less than the density of the water. ☑

Think about a hot air balloon, such as the one shown below. When the pilot heats the air in the balloon, the molecules move faster. They take up more space. The air in the balloon expands. The air inside the balloon is less dense than the air around the balloon. The balloon begins to rise and float.

Once the balloon is in the air, thermal energy flows from the balloon to the surrounding air. As the air in the balloon cools, its volume decreases and its density increases. When the balloon becomes denser than the surrounding air, it sinks back to the ground. It will take more heat energy for the balloon's density to decrease and keep it floating.

✔ Reading Check

3. **Summarize** What determines if an object will float in water?

_____

_____

_____

Picture This

4. **Identify** Circle the source of heat shown in the cutaway section of the second balloon.

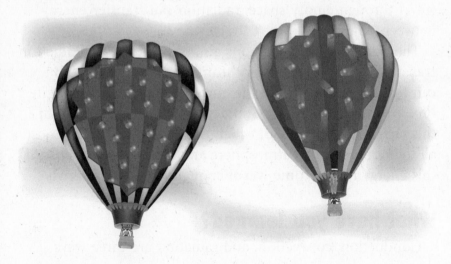

What effect does temperature have on fluid density?

Convection occurs in a fluid when an increase in temperature decreases the density of the fluid. The heated fluid begins to float upward because it is less dense than the surrounding fluid. For example, when the water in the bottom of a beaker is heated, its temperature increases. The average kinetic energy of the particles increases. The particles move faster and spread apart. The decreased density causes the warm water to rise.

What are convection currents?

When the warm water at the bottom of the heated beaker moves to the surface of the beaker, the particles lose kinetic energy by colliding with other particles. The particles move more slowly and get closer together. The density of water near the top of the beaker increases. It is pushed aside by rising warm water. The cooler, denser water at the top sinks along the sides of the beaker. The water may be heated again once it reaches the bottom of the beaker. This movement of water in the beaker is called a **convection current**. ☑

Radiation

Conduction and convection are forms of thermal energy transfer that occur only in matter. Another type of thermal energy transfer is radiation. **Radiation** (ray dee AY shun) is the transfer of thermal energy by electromagnetic waves. Remember that electromagnetic waves can travel in matter and in empty space. Because electromagnetic waves can travel through empty space, radiation can transfer energy between objects that aren't touching.

The Sun gives off <u>enormous</u> amounts of energy. Radiation transfers some of this energy from the Sun to Earth. You can feel this energy when the Sun shines on your skin. Life on Earth depends on the Sun's energy. Plants use some of this energy to make food, which provides energy for almost all living things. The atmosphere and Earth's surface absorb this energy and become warm enough for life to survive.

What have you learned?

Conduction, convection, and radiation are three ways that thermal energy moves from place to place. Conduction transfers thermal energy by collisions between particles in matter. Convection transfers thermal energy by the movement of matter from one place to another. Radiation transfers thermal energy by electromagnetic waves.

☑ Reading Check

5. Explain What happens to the water's density when its particles lose kinetic energy?

_____

_____

_____

Academic Vocabulary
enormous (ee NOR mus) (adj.)
marked by extraordinarily great size, number, or degree

💡 Think it Over

6. Compare Which type of energy transfer occurs when an ice cube melts in a cup of tea? (Circle your answer.)
   a. conduction
   b. convection
   c. radiation

Plate Tectonics

lesson ❶ Continental Drift

Grade Six Science Content Standard. 1.a. Students know evidence of plate tectonics is derived from the fit of the continents; the location of earthquakes, volcanoes, and midocean ridges; and the distribution of fossils, rock types, and ancient climatic zones.

● Before You Read

Think about the different continents on the surface of Earth, their shapes and the islands scattered throughout the oceans. Describe below how you think the continents gained their shape. Do you think that shape could ever change? Read on to find out more about Earth's continents.

_____

_____

_____

_____

MAIN ‹Idea

The hypothesis of continental drift was originally rejected by most scientists.

What You'll Learn
- why Alfred Wegener's continental drift hypothesis was so controversial
- evidence for continental drift
- about the ancient landmass Pangaea

● Read to Learn

Drifting Continents

Five hundred years ago, Europeans sailed across the Atlantic Ocean and found continents they had never seen before. Today, we call those continents North and South America. Mapmakers began to include them on their maps.

When mapmakers studied at the edges of North and South America, they became curious. The edges of the Americas looked like they might fit into the edges of Europe and Africa. The observation inspired Alfred Wegener to come up with a controversial idea.

What is continental drift?

If you look at a map, the edges of the continents bordering the Atlantic Ocean look kind of like pieces of a jigsaw puzzle. In the early 1900s a German scientist named Alfred Wegener decided the continents had been attached at one point. He proposed the theory of **continental drift**, the idea that the continents move slowly parallel to Earth's surface.

Mark the Text

Identify Main Ideas
Underline the main idea in each paragraph. Review the main ideas after you have read the lesson.

FOLDABLES™

Ⓐ Record Information
Make four note cards. Label the quarter sheets as illustrated and use them to record who, what, when and the arguments for continental drift.

Reading Essentials

Chapter 4 Plate Tectonics **37**

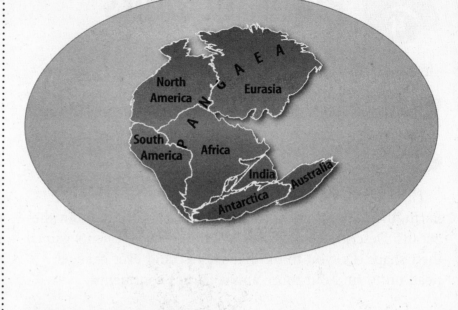

What was Pangaea?

Wegener thought that a few hundred million years ago, all the continents formed a huge landmass, which he named **Pangaea** (pan JEE uh), which is Greek for "all lands." He hypothesized that Pangaea existed around the time of the dinosaurs. Over millions of years, Pangaea would have broken and drifted apart to form the continents we recognize today, as shown above.

Evidence for Continental Drift

Wegener presented four types of evidence to support his continental drift <u>hypothesis</u>. This evidence included:
- how the continents fit together
- fossils
- rocks and mountain ranges
- ancient climate records

How did the continents fit together?

Wegener decided the continents would fit together if the Atlantic Ocean didn't separate them. For instance, the east coast of South America fits into the notch on the west coast of Africa. And, the bulge on northwest Africa fits right into the space between North and South America. Imagine the continents pieced back together like jigsaw puzzle pieces. Animals could have lived right over the space where the continents broke apart, and their fossils would be scattered on both sides of the break.

Academic Vocabulary

hypothesis (hi PAH thuh sus) (noun) an explanation for a problem that can be tested

Think it Over

2. Explain How does Pangaea add to what you know about the theory of continental drift?

_____

_____

_____

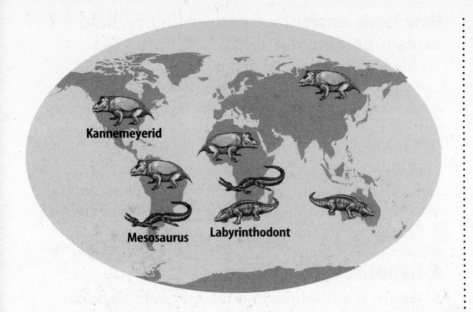

Kannemeyerid

Mesosaurus   Labyrinthodont

Picture This

3. Apply Fossil remains of the extinct plant *Glossopteris* have been found in Asia, Australia, Antarctica, Africa, and South America. Add a leaf symbol to these areas on the map to represent *Glossopteris*.

What does fossil evidence show scientists?

Fossils are the remains, imprints, or traces of once-living organisms. When an organism dies and gets buried in sediment, it can harden and be preserved. Remains in sediment that have hardened into sedimentary rock help scientists learn about past species, such as the dinosaurs shown above.

Wegener then studied a seed plant named *Glossopteris* (glahs AHP tur us) with fossils scattered through South America, Africa, India, Australia, and Antarctica. Such seeds couldn't have blown or floated across the large ocean that separates the continents. Wegener reasoned when the plant was alive those continents must have been attached.

How do rock types and mountain ranges explain continental drift?

Geologists can recognize rocks as groups because of their age and composition. Wegener discovered that distinctive groups of rocks on the continents would match up if the Atlantic Ocean were removed. For instance, specific rock groups in Africa match ones in South America, and indicate where the two continents would have been attached during the time of Pangaea. ☑

Certain mountain ranges on separate continents also connected together when Pangaea existed. For instance, Wegener realized the Appalachian Mountains in eastern North America could have matched similar mountain ranges in Greenland, Great Britain, and Scandinavia. Such evidence supports the idea of Pangaea.

☑ **Reading Check**

4. Explain What evidence from mountain ranges did Wegener use to support the theory of continental drift?

_____

_____

_____

_____

_____

_____

How does ancient climate evidence help explain continental drift?

Wegener traveled the world looking for rocks that contained evidence of past climates. When sedimentary rocks form, they preserve clues about the climate in which they formed. For instance, a hot, wet climate has many plants. Such plants form coal deposits in sedimentary rock. Similarly, a hot, dry climate produces rocks with preserved sand dunes. By looking at the patterns produced in sedimentary rock, Wegener determined that the climate during the time of Pangaea wasn't the same as the climate today.

A Hypothesis Rejected

Wegener presented this evidence for continental drift to other scientists. However, Wegener had difficulty explaining how, when, or why the continents slowly drifted across Earth's surface. He proposed that the continents drifted by plowing through the seafloor. He thought the same forces of gravity that produced tides in the ocean had moved the continents.

Most scientists did not accept Wegener's explanation. Because these scientists could not think of any forces strong enough to make continents drift, Wegener's hypothesis was rejected.

Continental Drift Hypothesis

Wegener proposed that the continents drifted by plowing through the seafloor. He thought the forces that make the tides and waves in the ocean also moved the continents.

Wegener's evidence included the geographic fit of the continents, fossils, rocks and mountain ranges, and ancient climate records. Because scientists could not think of forces strong enough to make continents drift, most scientists rejected Wegener's theory.

Picture This

6. Label Find North America on each of the three drawings. Label it.

How Continents May Have Drifted

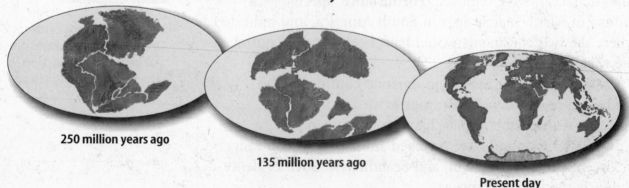

250 million years ago

135 million years ago

Present day

Science⌇nline ca6.msscience.com

Plate Tectonics

lesson ➋ Seafloor Spreading

 Grade Six Science Content Standard. 1.a. Students know evidence of plate tectonics is derived from the fit of the continents; the location of earthquakes, volcanoes, and midocean ridges; and the distribution of fossils, rock types, and ancient climatic zones.

⬤ Before You Read

Think about why Wegener's theory of continental drift was rejected. Describe on the lines below the kind of evidence that might have made others accept his ideas. Then read the lesson to find out what scientists learned through studying Earth's seafloor.

_____

_____

_____

⬤ Read to Learn

Investigating the Seafloor

Wegener studied Earth's land surface to prove the theory of continental drift. Other scientists looked at the seafloor. Most rocks in the seafloor have molecular compositions with a lot of basalt and little silica, a composition different from rocks on land. Starting in the 1950s, scientists developed new tools and technologies to study the seafloor. Those new tools and technologies helped scientists find out why seafloor rocks differed from rocks on land.

What is sonar?

During World War II, scientists developed a technology called sonar, which helped them map the seafloor. Sonar uses sound waves like a bat uses sound to navigate. When scientists emit soundwaves from a boat, the soundwaves hit the land at the bottom of the ocean and bounce back. Scientists know how fast sound travels in water. A receiver records how long it takes for sound to bounce off the ocean floor and return to the boat. The data helps scientists learn the ocean's depth, and map the seafloor.

MAIN ‹ Idea

Seafloor spreading is one explanation for continental drift.

What You'll Learn
- evidence for seafloor spreading
- how scientists determined the age of the seafloor

Study Coach

Define Key Terms Write down the underlined words in this lesson on a separate sheet of paper. Write a definition and use each term in a sentence.

FOLDABLES™

Ⓑ Record Information Make four note cards. Label the quarter sheets as illustrated and use them to record who, what, when, and the arguments for seafloor spreading.

What are mid-ocean ridges?

When scientists mapped the seafloor, they found the largest mountain ranges on Earth. These underwater mountain ranges wrap around Earth, and are called **mid-ocean ridges**. More heat comes out of Earth from mid-ocean ridges than anywhere else on Earth. ☑

The Seafloor Moves

Harry Hess, an American geologist who studied the seafloor, wanted to know what formed mid-ocean ridges. He thought the mid-ocean ridges became hot because lava erupted from them to make a new seafloor.

Hess called his theory **seafloor spreading**, the process by which mid-ocean ridges constantly make new seafloor. Earth's mantle rises towards the surface, and magma forms, as shown below. The magma comes out of cracks in the ridges as lava. When the lava cools, it forms a new seafloor. The seafloor cools and moves away from the mid-ocean ridge, as a new part of Earth's crust. This makes the ocean wider.

As the ocean widens, continents drift apart. Thus seafloor spreading could explain continental drift. Scientists looked for evidence that Hess's theory was correct.

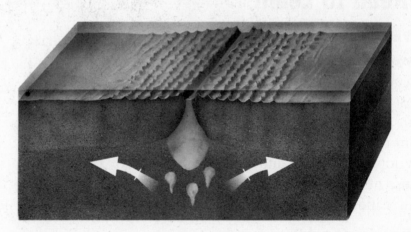

Evidence for Spreading

If Hess was right, the seafloor would be youngest at the mid-ocean ridge. Scientists discovered this was true, using Earth's magnetic field. When you hold a compass, the needle aligns to Earth's magnetic field. The north-seeking end of the needle always points north. This wasn't always true.

Picture This

2. Interpret Circle the mid-ocean ridge in the figure. Trace over the arrows that show the direction in which the old seafloor is moving.

Do magnetic poles reverse?

Earth's magnetic field hasn't always had the same orientation. Magnetic poles sometimes reverse. The process can take 10,000 years or millions of years. The next time the poles reverse, if you're still alive, your compass needle will point south, and not north.

Scientists use the term *normal* to refer to the magnetic field today. If one day south becomes north, scientists will call the magnetic field *reversed*. Scientists have documented each time the poles have reversed throughout time.

Igneous rocks can record when the magnetic poles <u>reverse</u>. This happens along a mid-ocean ridge when basalt forms from lava and cools. Tiny, iron-rich crystals harden inside the layers of rock and their resulting orientation forms a record of Earth's magnetic field from the time the rocks cool.

Why did scientists drill into the seafloor?

Soon after scientists learned how to find the age of the seafloor by looking at when Earth's magnetic poles reversed, they developed deep-sea drilling. Seafloor drilling began in 1968, when a boat called the *Glomar Challenger* drilled kilometer long pipes into rock at the bottom of the sea and brought up samples.

Seafloor drilling helped prove that the idea of seafloor spreading was correct. Rock samples showed the oldest rocks lay furthest from mid-ocean ridges. The youngest rocks lay right at the center of mid-ocean ridges. ☑

Seafloor Spreading Hypothesis

By the 1950s, new models and technologies, such as sonar, were being developed to map and explore the seafloor. When scientists mapped the topography of the seafloor they discovered underwater mountain ranges known as mid-ocean ridges. Harry Hess studied the seafloor trying to understand how mid-ocean ridges were formed. He proposed the seafloor spreading hypothesis, which is the process by which new seafloor is continuously made at the mid-ocean ridges. New evidence from around the world showed that the seafloor was spreading, as Hess had thought. Seafloor spreading seemed to explain continental drift. Studies of mid-ocean ridges continue today.

Academic Vocabulary
reverse (rih VURS) (verb) to completely change position or direction

☑ **Reading Check**

3. **Describe** Where do the oldest rocks in the seafloor lay in relation to the youngest rocks in the seafloor?

_____

_____

_____

_____

_____

Plate Tectonics

lesson ⊜ Theory of Plate Tectonics

Grade Six Science Content Standard. 1.c. Students know lithospheric plates the size of continents and oceans move at rates of centimeters per year in response to movements in the mantle. **Also covers:** 1.b, 4.c.

MAIN ⟨Idea

Earth's lithosphere moves as a result of different forces.

What You'll Learn

- what the theory of plate tectonics is, and how plate tectonics works
- evidence for plate tectonics
- the difference in oceanic and continental lithosphere

Study Coach ▶

Ask Questions As you read, write two questions for each page of this lesson. Discuss your questions and answers with classmates. If you cannot find the answer, ask your teacher for help.

FOLDABLES™

C Record Information
Make four note cards. Label the quarter sheets as illustrated and use them to record what you learn about earth's plates, plate boundaries, plate movement, and the rock cycle.

Earth's Plates	Plate Boundaries
Plate Movement	The Rock Cycle

● Before You Read

Many artists use tiny tiles to create works of art called mosaics. Describe below what would happen if you moved the pieces to different locations. Then read on to find out more about how Earth's surface fits together.

_____

_____

● Read to Learn

Earth's Plates

Seafloor spreading showed how the continents could move. They don't drift along the seafloor as Wegener thought. The continents and the seafloor move over a weaker layer in the mantle of the Earth.

Geologist J. Tuzo Wilson named the large pieces of Earth's surface that move across one another *plates*. Wilson said Earth's plates were like pieces of a splintered eggshell, outlined by cracks. He thought seven large plates covered Earth. Today, the large brittle pieces of Earth's outer shell, or surface, are called **lithospheric plates**.

What is the theory of plate tectonics?

The theory of **plate tectonics** explains how lithospheric plates move and cause major geologic features and events on Earth. The theory was founded on ideas like continental drift and seafloor spreading and is widely accepted today.

Plate boundaries are shown as thin lines on the map on the next page. Earthquakes and volcanoes are common at plate boundaries, such as those around the edges of the Pacific Ocean. Many volcanoes and earthquakes are located near long, deep parts of the seafloor called **ocean trenches**.

Major Plates of the Lithosphere

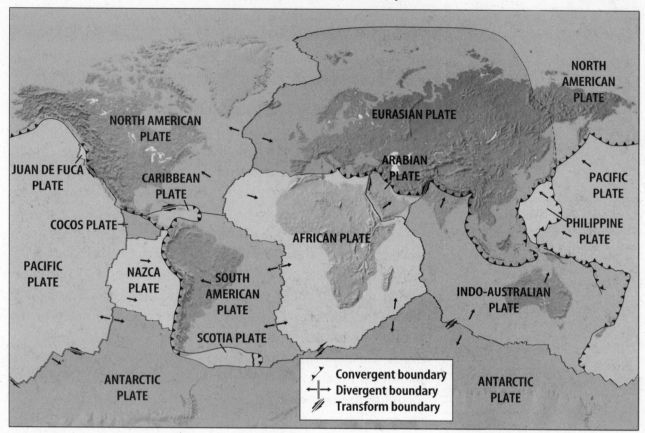

Convergent boundary
Divergent boundary
Transform boundary

What type of rocks are found in the lithosphere?

The lithosphere is made of Earth's crust and upper mantle. Rigid and brittle rocks are found in the lithosphere. When forces such as volcanoes or earthquakes act on these rocks, the rocks bend or break.

What's the difference between oceanic and continental lithosphere?

The two types of lithosphere are oceanic and continental. Most of the major plates have both oceanic lithosphere and continental lithosphere.

Oceanic lithosphere is much thinner than continental lithosphere. Oceanic lithosphere forms near a mid-ocean ridge. It contains oceanic crust, which is made mainly of dense basalt, gabbro, and a thin layer of sediment.

In contrast, continental lithosphere contains continental crust, which is made of igneous rocks like granite, metamorphic rocks like gneiss, and a covering of sedimentary rocks. Most of the rocks in the continental lithosphere are less dense than those in the oceanic lithosphere. ☑

Picture This

1. **Locate** Notice the small arrows on the figure above. Find and circle two arrows pointing in opposite directions. These arrows show two plates moving away from each other.

✔ Reading Check

2. **Compare** Which lithosphere is denser? (Circle your answer.)
   a. oceanic
   b. continental

What controls plate movement?

In the past, scientists thought forces deep within Earth called convection currents moved lithospheric plates above them. Now, new studies have made scientists think other forces might control the movement of lithospheric plates.

What causes convection?

Heat has been escaping from Earth since it first formed. One way heat transfers from Earth's deep interior, or core, to the surface is by convection. Convection is a type of heat transfer that occurs as a result of temperature differences in Earth's mantle. ☑

Convection requires a continuous supply of internal heat. One heat source comes from radioactive decay of elements like uranium in Earth's mantle. Heat and radiation get produced in rocks with radioactive elements. The heat lowers the density of the surrounding rock. The heated rock is pushed upward. This moves both rock and heat from inside Earth toward the surface. Convection currents provide matter and energy for the motion of the plates. The arrows in the figure below show the rise and fall of material deep inside Earth.

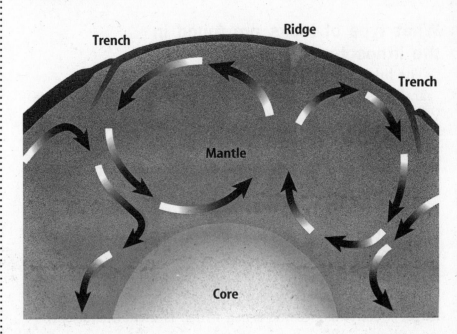

✔ Reading Check

3. Identify What is the Earth's interior called?

Picture This

4. Summarize Choose one convection current in the figure and label the arrows *heating*, *rising*, *cooling*, and *sinking*.

How might convection drive plate movement?

At a mid-ocean ridge, less dense rock is brought to Earth's surface. Because the surrounding pressure deep within Earth is so great, the rock melts into lava. The lava cools at the surface as a new part of an oceanic ridge. When a new convection current occurs, the new rock, formed from the lava, moves away from the ridge, and begins to cool. As the rock cools, it becomes more dense.

As convection currents rise to Earth's surface, they move the brittle plates of Earth's lithosphere. The plates move over Earth much like ice floats on top of water. Sometimes a cooler, more dense mass of rock sinks down into the mantle. Scientists call these sinking rocks **slabs**. When a slab sinks, it bends and breaks. This causes an earthquake. ☑

What are ridge push and slab pull?

Ridge push and slab pull are two forces that might be important in controlling the movement of plates. A mid-ocean ridge is higher than most of the rest of the seafloor. So when a plate rises to the top of a ridge, gravity helps push it down the ridge's slope, a process called ridge push.

Slab pull occurs when one plate gets forced into the mantle. As a cool, dense slab sinks into the mantle, gravitational forces act on the dense slab and pull it even deeper into the mantle.

Scientists now study plates, mid-ocean ridges, slabs, and complex waves that travel through Earth to learn what forces control the movement of plates.

Measuring Plate Movement

Since the 1960s, scientists have found new ways to prove the theory of plate tectonics. Now they can measure plate movement directly, with a system of satellites.

What is the Global Positioning System?

The **Global Positioning System** (GPS) is a network of satellites used to determine locations on Earth. A receiver on Earth processes signals from several satellites circling the planet. A computer in the receiver calculates the exact coordinates of the receiver's location on Earth. Today, you can find GPS in cars, planes, and even in handheld devices.

Scientists use GPS to measure plate movements. Certain receivers monitor the direction and speed of plates as they move across the planet.

☑ Reading Check

5. Explain What happens when a slab sinks?

_____

_____

☼ Think it Over

6. Draw Conclusions Name another task a GPS might be useful for?

_____

_____

_____

_____

What is satellite laser ranging?

Satellite laser ranging (SLR) uses pulses of light to measure distances. These pulses of light are laser beams. Measurements using SLR show similar results to those made using GPS. The results from the two systems show us that plates move about as fast as your fingernails grow, only centimeters a year.

Plate Tectonics and the Rock Cycle

Plate tectonics make the rock cycle easier to understand. Rocks constantly move through the rock cycle. Magma rises and becomes igneous rock at the mid-ocean ridge. A plate moves away from the ridge slowly and cools, carrying igneous rock along with it. Eventually the plate is pushed back into the mantle and part of it may melt. So while some rocks <u>erode</u> and melt on mountaintops, others form at the same time on the seafloor. The rock cycle occurs slowly, over millions of years. Rocks are constantly recycled as the rock cycle continues. ☑

Plate Movements

Earth's outer shell is broken into large, brittle pieces called lithospheric plates. The thickness and composition of a lithospheric plate varies. The thickness and composition depends upon whether the plate is made of oceanic or continental material. Forces in Earth, such as earthquakes and volcanoes, occur along the boundaries of the lithospheric plates.

The theory of plate tectonics explains how lithospheric plates move and cause major geologic features and events on Earth's surface. Some scientists hypothesize that convection currents in Earth's mantle drive the movement of plates. And, some scientists think that ridge push and slab pull are two forces that control the movement of plates.

Academic Vocabulary

erode (ee ROHD) (verb) to wear away over time

✔ **Reading Check**

7. **Determine** What happens in the rock cycle after magma rises to the surface of an ocean ridge and becomes igneous rock?

_____

_____

_____

_____

Plate Boundaries and California

lesson ❶ Interactions at Plate Boundaries

Grade Six Science Content Standard. 1.d. Students know that earthquakes are sudden motions along brakes in the crust called faults and that volcanoes and fissures are locations where magma reaches the surface. **Also covers:** 1.c, 1.e.

⬤ Before You Read

Have you ever experienced an earthquake? Write a paragraph describing your experience. If you haven't experienced an earthquake, use your imagination to describe how you would stay safe during an earthquake. Read the lesson to learn about the types of stresses that cause earthquakes.

_____

_____

_____

_____

MAIN ⟨Idea

At plate boundaries stresses cause rocks to deform.

What You'll Learn

- types of stresses that deform rocks
- how geologic features of Earth's surface indicate a type of plate boundary
- how subduction zones relate to volcanoes

⬤ Read to Learn

Stress and Deformation

Earth's lithosphere—the crust and part of the mantle—is broken into plates, packed closely together. The plates move in different speeds and different directions, so stress and collisions occur. The plates move slowly, but their size means collisions are extremely powerful.

Can rocks bend?

Rocks can bend without breaking. Rocks stressed at high temperatures may change shape by folding, but not breaking. Scientists call this plastic deformation. When stress occurs slowly, or at high temperatures, these deformations are likely to occur. Rocks sometimes snap back to their original shape when stress no longer exists. This is called elastic deformation. ☑

A break in a rock is called a **fracture**. If rocks on one side of a fracture have moved relative to the rocks on the other side, the fracture is called a **fault**.

▷ Study Coach

Answer Questions After you have read the information below each question heading, write the answer to the question in your own words on a separate sheet of paper.

✔ Reading Check

1. Explain What is elastic deformation?

_____

_____

_____

_____

What three main types of stress lead to faulting?

Three types of stress that lead to faulting include tension, compression, and shear stress. In nature, combinations of these stresses often occur. Each type of stress can cause more than one type of fault.

What is tension stress?

When two plates move apart, as shown in the figure below, the lithosphere must stretch and become thinner. The result is a deformation caused by tension.

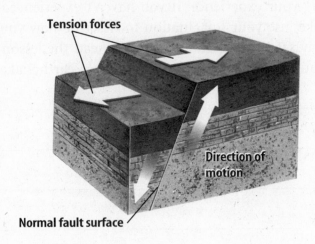

Tension forces

Direction of motion

Normal fault surface

What is compression?

When plates are squeezed together, as shown below, the rocks become thicker. This deformation is called compression.

Compression forces

Direction of motion

Reverse fault surface

What is shear stress?

Rocks can also be pushed sideways in opposite directions. When this occurs, the stress is called shear. Rocks do not become thinner or thicker from shear stress.

Picture This

2. Determine Which of the following is the result of tension stress? (Circle your answer.)
   a. rocks are pulled apart
   b. rocks are pushed together

Picture This

3. Compare What is the main difference between the two figures on this page?

_____

_____

What do geologists take note of when examining a fault?

Geologists examine faults to determine which way broken sections of the rock have moved. They look for objects broken by the fault to see what stresses caused movement.

Imagine a fault in which the rocks have been pulled apart. Geologists call the upper surface, from which you could hang, the hanging wall. The block of rock below the fault is called the footwall.

What types of faults do geologists identify?

Rocks along faults can move up, down, or sideways. Tension stress produces <u>normal</u> faults. The fault at Death Valley is a normal fault. Its hanging wall has moved down to the footwall.

Compression produces reverse faults. As compression pushes rocks together, the hanging wall moves up relative to the footwall. ☑

Shear stress produces strike-slip faults. Strike-slip faults are often vertical, meaning the rocks scrape sideways, or horizontally, past one another rather than moving up or down. The San Andreas fault is a strike-slip fault.

Types of Plate Boundaries

Earth's plates meet at plate boundaries. Faults form along these plate boundaries. Three types of plate boundaries exist. The type of boundary is determined by the way the rocks on either side of the boundary move. Geologic features formed at faults depend on the type of plate boundary and type of stresses generated.

What happens at a divergent plate boundary?

Geologists use the term **divergent plate boundary** when lithospheric plates are moving or pulling apart. Seafloor spreading helps create mid-ocean ridges at divergent plate boundaries.

Mid-Ocean Ridges As lithospheric plates move apart at mid-ocean ridges, new seafloor forms, rocks break, and earthquakes occur. Rocks cool and contract becoming denser as they move away from a mid-ocean ridge. As a result, the seafloor farther from the ridge is deeper underwater than the seafloor near the ridge. Studying mid-ocean ridges is difficult because the ridges are located about 2 km below sea level.

Academic Vocabulary
normal (NOR muhl) (adj)
occurring naturally; not unusual

✔ **Reading Check**

4. Identify What kind of stress produces reverse faults?

_____

_____

FOLDABLES™

Ⓐ **Sketch and Describe**
Make a layered Foldable. Label the tabs as illustrated. Under the tabs, record main ideas, vocabulary terms, definitions, and sketch examples of each type of plate boundary.

Plate Boundaries
Divergent Boundaries
Convergent Boundaries
Transform Boundaries

Continents Pull Apart Most divergent plate boundaries are located on the seafloor. Some divergent plate boundaries form on land and pull continents apart. The process that pulls continents apart is called **continental rifting**.

When the lithosphere stretches, tension stress occurs, causing rocks to pull apart, break, and form normal faults. As the hanging wall of a normal fault slides down the footwall, a long, flat, narrow **rift valley** forms. ☑

If the pulling continues, the rift valley grows. If the valley reaches a shoreline, ocean water floods into it. In time, the gap between two plates might widen to form an ocean.

What happens at a convergent plate boundary?

A **convergent plate boundary** is formed when two plates move toward each other. Eventually, the plates collide. The results of the collisions depends on what the plates are made of—continental or oceanic lithosphere. When plates collide, earthquakes occur, and new topographic features form.

Ocean-to-Ocean Collision When two oceanic plates collide, a slab of one plate slides under the other, in a process called **subduction**. The density of the plate determines which plate subducts. Generally, a colder, denser slab is forced down into the mantle and a trench forms.

As the slab sinks, temperature and pressure release water from minerals in the slab. Where the water rises into the mantle, rocks melt and create the magma that sprays out of volcanoes in the overlying plate, as shown in the illustration below.

Picture This
6. Interpret Circle the plate that is going underneath the other plate.

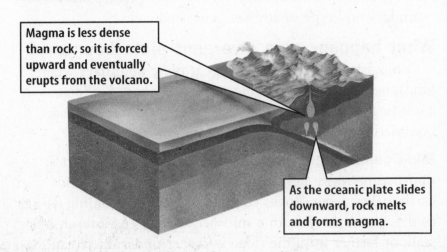

Magma is less dense than rock, so it is forced upward and eventually erupts from the volcano.

As the oceanic plate slides downward, rock melts and forms magma.

Ocean-to-Continent Collision When an oceanic plate collides with a continental plate, the oceanic plate always subducts. The subducting plate creates a curved line of volcanoes along the edge of the overlying continental plate.

Continent-to-Continent Collision When two continental plates collide, neither subducts. The granite and shale rocks of continental plates aren't dense enough to sink. Instead, compression stresses force uplift. When continents collide and are uplifted, tall mountains form. ☑

What are transform plate boundaries?

When two plates scrape sideways past one another, a **transform plate boundary** exists. Lithosphere isn't formed or recycled at these boundaries. A transform plate boundary is similar to a strike-slip fault. Plates scrape past each other and slip, rocks break, and earthquakes occur.

Oceanic Oceanic transform plate boundaries connect pieces of mid-ocean ridges. Mid-ocean ranges are often made of many short pieces. In fact, most oceanic transform plate boundaries are short.

Continental Some transform plate boundaries occur in continental lithosphere. Large earthquakes occur along the faults. The San Andreas Fault in California is a continental transform boundary. The boundary separates the North American Plate and the Pacific Plate. ☑

Deformation and Plate Boundaries

Scientists study how rocks break and bend. They identify the stresses that cause deformation such as fractures and faults. They use clues from these studies to unravel Earth's geologic history. Scientists use data to find out the direction and distance plates have moved, and to determine how plates have interacted at plate boundaries.

☑ **Reading Check**

7. **Explain** When two continental plates collide, why doesn't one subduct?

_____

_____

_____

☑ **Reading Check**

8. **Determine** What do continental transform boundary regions separate?

_____

_____

_____

Plate Boundaries and California

lesson ❷ California Geology

 Grade Six Science Content Standard. 1.f. Students know how to explain major features of California geology (including mountains, faults, volcanoes) in terms of plate tectonics. **Also covers:** 1.e.

MAIN ‹Idea

Many California landforms were produced by plate tectonic activity.

What You'll Learn

- how plate movements have produced California's landforms
- how California's topography might change in the future

Study Coach ▶

Make an Outline As you read, make an outline of the lesson. Use the headings for your outline structure.

Picture This

1. **Identify** The top of the image is north. What directions does the Pacific Plate move in relation to the North American Plate?
   a. south and east
   b. north and west

● Before You Read

Describe some of California's geologic features on the lines below. Then read the lesson to learn about how California's landforms were formed.

_____

_____

● Read to Learn ·······················

Plate Tectonics in California

A continental transform boundary cuts across California. At the northern end of the state, a convergent plate boundary sits offshore. The movement of these plates produces earthquakes, volcanoes, and mountains.

What is California's transform plate boundary?

Most of California is located on the North American Plate. A small part of California lies on the Pacific Plate. The plate does not slide smoothly, but moves in jerks, which causes earthquakes. The <u>San Andreas Fault</u> is the transform plate boundary between these two plates, as seen below.

The Pacific Plate moves northwest relative to the North American Plate. The velocity of movement is about 3.4 centimeters per year.

San Andreas Fault · North American plate · San Francisco Bay · San Francisco · Pacific plate

Bends in the Fault The San Andreas Fault is not a straight line. Bends in the fault mean blocks of rock have been pushed up or down, forming mountains or basins. Many of California's mountains formed because the Pacific Plate moves past the North American Plate. The Los Angeles Basin, the Ventura Basin, and San Francisco Bay formed as blocks of rock have dropped down. ☑

What forms California's convergent plate boundaries?

Just offshore of Northern California are two small oceanic plates—Gorda and Juan de Fuca. These plates are subducted beneath the coast at the Cascadia Subduction Zone. The subduction forms a convergent plate boundary. The volcanic mountains of California's Cascade Range formed at this subduction zone.

California's Mountains

California's numerous mountain ranges often formed from the interactions at several plate boundaries. As rocks on one side of a transform plate boundary scrape and push against the rocks on the other plate, mountains, such as the Transverse Ranges, can form.

How does subduction form California's geographic features?

California's convergent plate boundaries, in the past and the present, have been important in forming California's mountains. Rocks made of granite form under volcanic mountains where plates meet. And during mountain building, forces that compress and increase heat produce metamorphic rocks. For example, the Klamath Mountains and Sierra Nevada contain igneous and metamorphic rocks that formed below the surface. These rocks formed when an ancient oceanic plate subducted beneath the North American Plate. ☑

In Northern California, granite rocks are forming deep in the crust. Volcanic activity on the continental lithosphere above the subduction zone in northern California produces the Cascade Range. Lassen Peak and Mount Shasta in the Cascade Range are active volcanoes. The figure at the top of the next page shows a subduction zone and what can occur at a convergent boundary between oceanic and continental plates.

FOLDABLES

B **Draw** Make a two-tab Foldable. Label the tabs as illustrated. Sketch a map of California on top of the front tabs. Under the tabs, draw and label the San Andreas Fault line and write about the cause and effect of the San Andreas Fault.

Cause

Effect

Reading Check

2. Determine What kind of forces produce metamorphic rocks?

_____

_____

Picture This

3. Explain As the oceanic plate sinks into the mantle, it melts and gets smaller. What is happening to the continental plate that balances this action?

_____

_____

_____

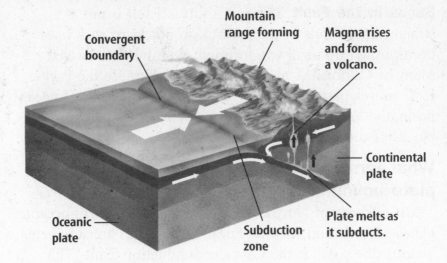

Convergent boundary

Mountain range forming

Magma rises and forms a volcano.

Continental plate

Oceanic plate

Subduction zone

Plate melts as it subducts.

What mountains did continental rifting produce?

Tension stress has also formed some mountains in California. The Panamint Range west of Death Valley is rising as the crust in eastern California stretches. Mountains help scientists understand processes that are part of California's tectonic history.

Future Plate Movement

Scientists can <u>estimate</u> the directions and speeds plates move. They can predict future plate boundary interactions. The part of California that is on the Pacific Plate, including Los Angeles, will continue to move northwest along the coast relative to the North American Plate. This means Los Angeles and San Francisco are approaching each other about as fast as your fingernail grows, forming mountains and basins along the way. ☑

Academic Vocabulary

estimate (ES tuh mayt) (verb) to make an educated guess or calculation

☑ **Reading Check**

4. Explain How can scientists predict future plate movements?

_____

_____

_____

Science Online   ca6.msscience.com

chapter 6 Earthquakes

lesson ❶ Earthquakes and Plate Boundaries

Grade Six Science Content Standard. 1.d. Students know that earthquakes are sudden motions along breaks in the crust called faults and that volcanoes and fissures are locations where magma reaches the surface. **Also covers:** 1.e, 1.f.

● Before You Read

Have you ever felt an earthquake? Or, was there ever a time when you felt unsteady, like the ground was going to slip out from under you? Write a description of your experience below. Read the lesson to learn about how earthquakes occur.

_____

_____

MAIN ‹ Idea

Most earthquakes happen at plate boundaries.

What You'll Learn
- what an earthquake is
- how earthquakes and faults are related

● Read to Learn

What is an earthquake?

An **earthquake** is the rupture and sudden movement of rocks along a fault. A fault is a fracture along which rocks can slip. Earthquakes occur when a fault ruptures, or breaks, when rocks are stretched so much that they can no longer stretch or flow. The rupture causes release of complex waves.

Most earthquakes happen in Earth's crust. Some occur where lithospheric plates subduct, where plates get recycled back into the Earth. But large earthquakes can occur far from plate boundaries.

How can heat within Earth lead to an earthquake?

Stress at the boundaries of Earth's lithospheric plates sometimes cause breaks and rock movements. Heat in Earth's mantle provides energy for this plate movement. Some heat energy inside Earth changes into kinetic, or moving, energy. ☑

Kinetic energy is transferred to rocks near faults. When plates can't release their kinetic energy, rocks are stressed and reach their elastic limit. The energy stored as a change in shape is called **elastic strain**. When rocks can no longer change shape, they eventually break and cause earthquakes.

Mark the Text

Highlight Definitions
Highlight or underline the definition of each of the key terms in this lesson. Ask your teacher to explain any key terms you do not understand.

✔ Reading Check

1. Define What is kinetic energy?

_____

How do earthquakes occur at faults?

Elastic strain energy builds up along a fault as rocks move past each other. The elastic strain is partly released when rocks break along the fault as shown below. This breaking sends complex waves into the surrounding rocks. The energy in the waves causes the shakes during and after earthquakes.

Picture This

2. Identify Trace over the arrows in the figure that show the direction that rock moves in a fault.

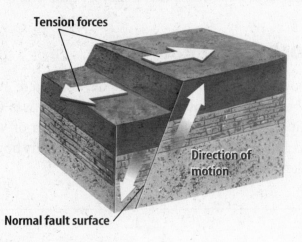

Tension forces

Direction of motion

Normal fault surface

What is the focus?

Earthquakes start at the **focus**, the location on a fault where rupture and movement begin. As the rupture gets bigger, it releases more energy into surrounding rocks. How much we feel an earthquake depends on how close the earthquake is to Earth's surface, and how large the rupture is. Large faults can have larger ruptures and tend to produce large earthquakes.

What are fault zones?

A plate boundary is often shown as a single line on a map. Plate boundaries are actually much more complex than a single fault. Plate boundaries are zones of faulting that are 40 km to 200 km wide. For example, the San Andreas fault shown below is a group of faults that result from plate motion between the Pacific Plate and North American Plate.

Academic Vocabulary

complex (KAHM pleks)
(noun) made up of many or complicated parts

Picture This

3. Highlight the arrows showing that each plate is moving in the same direction.

San Andreas Fault

San Francisco

San Francisco Bay

North American plate

Pacific plate

Major Plates of the Lithosphere

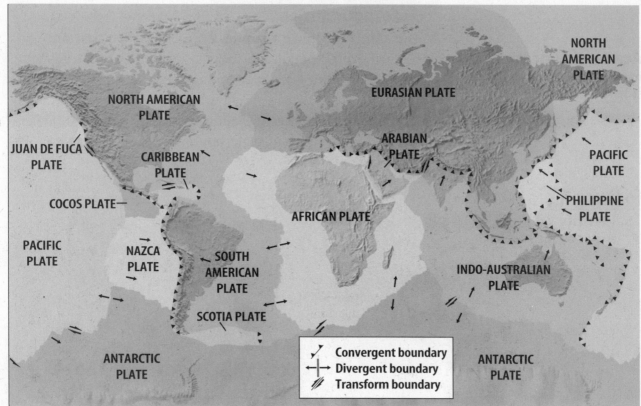

NORTH AMERICAN PLATE

EURASIAN PLATE

NORTH AMERICAN PLATE

JUAN DE FUCA PLATE

CARIBBEAN PLATE

ARABIAN PLATE

PACIFIC PLATE

COCOS PLATE

PHILIPPINE PLATE

PACIFIC PLATE

NAZCA PLATE

AFRICAN PLATE

SOUTH AMERICAN PLATE

INDO-AUSTRALIAN PLATE

SCOTIA PLATE

ANTARCTIC PLATE

ANTARCTIC PLATE

↗ Convergent boundary
⊢⊣ Divergent boundary
⫽ Transform boundary

Plate Boundaries and Earthquakes

There are patterns in earthquake size, earthquake depth, and the type of faults on which earthquakes occur. Plates interact at the different types of plate boundaries, shown in the figure above. Not all earthquakes happen at plate boundaries.

What happens at divergent plate boundaries?

As divergent plate boundaries move apart, tension stress causes rocks to break and form normal faults. Most earthquakes along these boundaries occur at shallow depths in the crust. The earthquakes at these places also tend to be small in size.

What happens at convergent plate boundaries?

Rocks at convergent plate boundaries get squeezed together. The deepest earthquakes occur at convergent plate boundaries. The deepest earthquakes happen at subduction zones. Earthquakes are deep at subduction zones because cool, rigid slabs sink deep into the warm mantle. The most devastating earthquakes recorded so far are associated with convergent plate boundaries.

Picture This

4. Highlight one boundary where Earth's plates converge, or come together.

![FOLDABLES]

A Describe Make a four-tab Foldable and label the front tabs as illustrated. Describe the earthquakes that occur at each of the boundaries under the tabs.

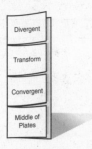

Divergent

Transform

Convergent

Middle of Plates

Picture This
5. Identify Circle the arrows on the strike-slip faults.

Think it Over

6. Predict How could preparing for an earthquake make it less dangerous?

_____

_____

_____

_____

_____

What are earthquakes like at transform plate boundaries?

At transform plate boundaries, rocks scrape past one another almost <u>parallel</u> to Earth's surface. The plates grind past one another and an earthquake can occur at the strike-slip faults, as shown in the figure below. These earthquakes also happen in relatively shallow depths. Major earthquakes occur where transform plate boundaries run through a continent.

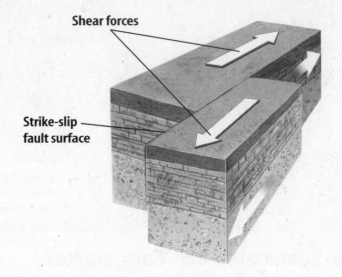

Shear forces

Strike-slip fault surface

Do earthquakes occur away from plate boundaries?

Most earthquakes occur along plate boundaries, where plates are moving. However, some earthquakes occur away from plate boundaries. Many occur in the middle of continents. These earthquakes don't occur often but can be quite dangerous, mainly because people don't expect them or prepare for them.

Causes of Earthquakes

Rocks can change shape and even snap back to shape if stresses are removed. Faults are created at plate boundaries where stresses causes rocks to change shape or become strained. Faults can rupture and move as earthquakes. However, some earthquakes also occur along faults located in the middle of plates. Seismic waves release some of the elastic strain energy stored in rocks.

Earthquakes

lesson ❷ Earthquakes and Seismic Waves

Grade Six Science Content Standard. 1.g. Students know how to determine the epicenter of an earthquake and know that the effects of an earthquake on any region vary, depending on the size of the earthquake, the distance of the region from the epicenter, the local geology, and the type of construction in the region.

● Before You Read

Sometimes, energy travels in waves. Think about the waves on an ocean beach, or the vibrating waves of motion when you pluck a guitar string. Write a paragraph on the lines below about what you think surfing on an ocean wave feels like. Then read the lesson that follows to learn about earthquake waves.

_____

_____

_____

MAIN ‹Idea

Earthquakes cause seismic waves.

What You'll Learn
- how energy from earthquakes travels in seismic waves
- the difference between primary, secondary, and surface waves
- how scientists investigate Earth's interior

● Read to Learn

What are seismic waves?

During an earthquake, the ground shakes, vibrates, and moves in all directions. In large earthquakes, people have actually seen waves on Earth's surface, rippling like water waves. Such waves of energy, produced at the focus of an earthquake are called <u>seismic</u> (SIZE mihk) <u>waves</u>. A seismologist is a scientist who studies earthquakes and seismic waves.

How do seismic waves travel?

Elastic strain energy builds up until it reaches the strength of the rock. Then the fault ruptures and releases some of the energy as seismic waves. Seismic waves move away from the focus in all directions. They travel to Earth's surface and toward Earth's core.

An earthquake's <u>epicenter</u> (EH pih sen tur) is the point on Earth's surface directly above the earthquake focus. Rocks absorb some energy as the waves move away from the focus and epicenter. The amount of energy decreases as waves move farther from the focus.

◀ **Study Coach**

Identify Main Ideas As you read, identify main ideas and concepts. Write down five main concepts and discuss them with a friend.

💡 **Think it Over**

1. **Compare** How are an earthquake's focus and epicenter related?

   _____

   _____

   _____

   _____

   _____

Picture This

2. Highlight the type of wave that causes the most damage.

Types of Seismic Waves

Earthquakes cause three different types of seismic waves: primary waves, secondary waves, and surface waves. Each travels differently within Earth.

What are primary waves (P-waves)?

Primary waves, also called P-waves, are compression waves. When a P-wave moves through rock, particles in the rock move back and forth parallel to the direction the wave travels. The energy moves by compressing and expanding the material through which it travels. Primary waves are the fastest seismic waves.

What are secondary waves (S-waves)?

Secondary waves, also called S-waves or shear waves, travel at right angles to the material they move through. The shearing movement changes the shapes of rocks. S-waves travel at about 60 percent of the speed of P-waves.

What are surface waves?

When P-waves and S-waves hit Earth's surface, energy can get trapped in the upper few kilometers of Earth's crust. This energy creates waves that travel along the surface. Surface waves travel more slowly than secondary waves. Rock particles can move side-to-side or have a rolling motion.

Surface waves often cause the most damage. They shake the crust more strongly than P-waves or S-waves. Their strong shaking can tear down buildings and break bridges. The table below summarizes the types of seismic waves.

Types of Seismic Waves	
Seismic Wave	**Description**
Primary waves (P-waves)	• cause rock particles to vibrate in same direction that wave travels • fastest seismic wave • first waves to be detected and recorded • travel through both solids and fluids
Secondary waves (S-waves)	• cause rock particles to vibrate perpendicular to direction that waves travel • slower than P-waves • detected and recorded after P-waves • travel only through solids
Surface waves	• cause rock particles to move with a side-to-side swaying motion or rolling motion • slowest seismic wave • generally causes the most damage

Using Seismic Wave Data

You have already learned that seismic waves travel at different speeds. Some seismic waves travel through Earth's interior and some travel along Earth's surface. Seismologists who study earthquakes and seismic waves, use this information to learn about Earth's interior.

What are the speeds of seismic waves?

All types of seismic waves start at the same time. Close to the focus of an earthquake, the S-wave is not very far behind the P-wave. But far from the focus, the S-wave travels far behind the P-wave.

Remember that seismic waves travel in all directions. If you stand in one position on the globe, P-waves will reach you first, followed by S-waves. Surface waves will be the last to reach you.

How do scientists map Earth's interior?

The speed and direction of seismic waves change when properties of Earth materials they travel through change. The waves bend or bounce off a new <u>layer</u> of rock. Scientists can track the patterns of primary and secondary waves to understand more about the rock layers in Earth's interior. As depth and pressure increases inside the Earth, rock layers become more dense. Seismic waves in the Earth curve when they hit these dense layers of rock.

What is the shadow zone?

Some large areas of the Earth don't receive any seismic waves from an earthquake. Seismologists call this area the shadow zone. In the shadow zone, S-waves stop and P-waves bend. S-waves travel only through solids. They stop when they reach Earth's outer core. When P-waves reach the outer core, their paths bend. The action of P-waves and S-waves also makes scientists think that Earth's outer core is liquid. ☑

Seismic Waves

Seismic waves shake the ground violently, causing destruction. However, without seismic waves from earthquakes, scientists would not have been able to measure the internal layering of the planet. How a seismic wave travels depends on the materials it travels through.

Think it Over

3. Draw Conclusions
If you feel a P-wave, followed quickly by an S-wave, which of the following is true?
(Circle your choice.)
a. You are near the earthquake's epicenter.
b. You are far from the earthquake's epicenter.

Academic Vocabulary

layer (LAY ur) (noun) one thickness of a substance lying over or under another

✔ Reading Check

4. Explain What happens when secondary waves reach the shadow zone?

_____

Earthquakes

lesson ❸ Measuring Earthquakes

Grade Six Science Content Standard. 1.g. Students know how to determine the epicenter of an earthquake and know that the effects of an earthquake on any region vary, depending on the size of the earthquake, the distance of the region from the epicenter, the local geology, and the type of construction in the region.

MAIN ‹Idea›

Scientists measure data from seismic waves to determine the location and size of an earthquake.

What You'll Learn

- what a seismograph does
- how scientists find an earthquake's epicenter
- different ways scientists measure earthquakes

Study Coach

Answer Questions For each question heading in this lesson, write your answer to the question on a separate sheet of paper.

✔ **Reading Check**

1. State Where did the December 26, 2004 water surge hit land?

_____

_____

● Before You Read

How could you measure how high a friend was jumping on a trampoline? How high could your friend jump if you were jumping too? How would you measure the difference? Write your thoughts below. Then read the lesson to find out how scientists measure earthquake waves.

_____

_____

_____

_____

● Read to Learn

How are earthquakes measured?

You might have read or heard about the earthquake and tsunami in the Indian Ocean on December 26, 2004. Scientists described the earthquake that caused the tsunami as having a magnitude of 9.0. What does this number mean? How do seismologists measure the size of an earthquake?

Compared to most other earthquakes scientists have recorded, the earthquake in the Indian Ocean in December 2004 was large. Scientists measured its size by determining how far the rock slipped along the fault. Scientists also measured the heights of the seismic waves recorded in that area. The seismic waves tell seismologists how much energy the earthquake released. Because the earthquake happened under water, the movement of the rock caused a wave to spread in the Indian Ocean. The wave grew in force until a giant surge of water reached the shores of South Asia and East Africa, killing hundreds of thousands of people and destroying the homes and businesses of more than a million people. ✔

Recording Seismic Waves

A **seismograph** (SIZE muh graf) is an instrument used to record and measure ground movements caused by seismic waves. It records the size, direction, and time of the movement. It includes information about the time each P-wave and S-wave arrives at a certain point.

How do seismographs record information?

Mechanical seismographs, such as those in the figure below, have a pen attached to a heavy weight, or pendulum, sitting over a drum covered with paper. When waves shake the ground, the drum moves. The pen and pendulum stay still. The moving drum lets the pen record the motion on paper wrapped around the drum.

Some seismographs record back-and-forth ground motion. Others record movement up-and-down. The paper record of such waves is called a **seismogram** (SIZE muh gram). With seismograms, scientists can calculate the size of earthquakes and determine the location of their epicenters.

How do you read a seismogram?

The horizontal, or flat, axis of a seismogram represents time. When the first primary seismic wave arrives and shakes the ground, it makes a wavy line on the record as shown below. Then a secondary wave arrives and makes another wavy line. Finally, big surface waves arrive. The seismogram shows how much each surface wave makes the ground move relative to other big surface waves. If the seismograph is near an earthquake, surface waves look bigger than if the seismograph is far away from an earthquake.

FOLDABLES

C **Explain** Make two note cards. Label two of the quarter sheets as illustrated and use them to record information on seismographs and seismograms. Explain the importance of each.

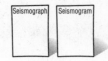

Picture This

2. Explain Why are two different seismographs used to record Earth's movements?

_____

_____

_____

_____

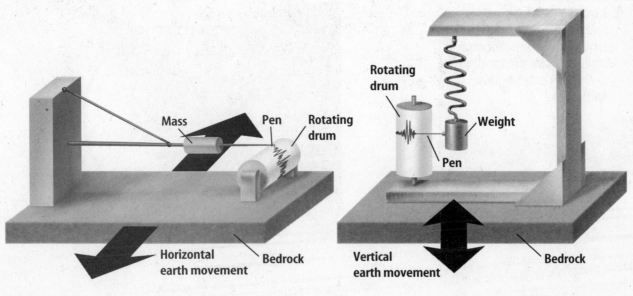

Mass   Pen   Rotating drum

Horizontal earth movement   Bedrock

Rotating drum   Weight   Pen

Vertical earth movement   Bedrock

How do scientists know how far primary and secondary waves travel?

Seismologists know the average speeds of primary and secondary waves. They can compare that information with a seismogram, which tells when waves arrived, to find out how far waves traveled.

Locating an Epicenter

Seismologists use the speeds of primary and secondary waves to help them find an epicenter. If three seismographs record waves from the same earthquake, the epicenter can be determined by a mathematical method called triangulation. ☑

What is triangulation?

To triangulate, scientists find the difference between the time primary and secondary waves arrive at a seismograph. Because seismologists know how fast primary and secondary waves travel, they use the time lag to find the distance to the epicenter. These calculations are performed individually for each seismograph recording of the earthquake.

At this point seismologists know how far the epicenter is from each seismograph. But they don't know in what direction the epicenter is located. That's where triangulation helps. A circle is drawn on a map around the location of each seismograph, with a radius equal to the distance the earthquake was from the seismograph, as shown below. The epicenter is located at the point where all the circles meet.

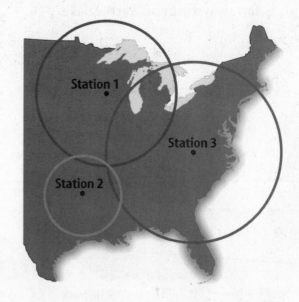

☑ **Reading Check**

3. **Identify** What mathematical method can help scientists find an epicenter?

Picture This

4. **Locate** Look at the figure with circles drawn around three seismograph stations. Mark the location of the epicenter of this earthquake.

Station 1

Station 3

Station 2

Measuring Earthquake Size

Seismologists use different types of scales to describe the size of earthquakes. Each scale uses different measurements to compare the size of one earthquake to another earthquake. Three examples of earthquake scales are described below.

What is the magnitude scale?

One scale, called the magnitude scale, is based on the amount of ground motion, measured by the height of seismic waves on a seismogram. A seismogram is shown below. Bigger earthquakes make bigger waves on a seismogram. Magnitude values range between 0 and 9. Each increase in number indicates a tenfold increase in ground shaking and 30 times more energy released.

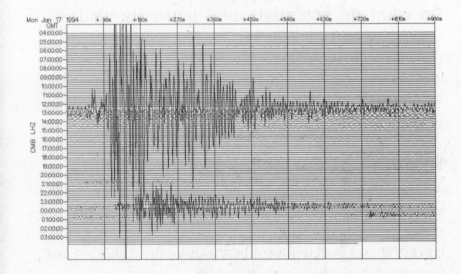

Picture This

5. Interpret Circle the area on the seismogram that recorded the earthquake.

What is the Richter magnitude scale?

Charles Richter designed the Richter magnitude scale for use in southern California in 1935. All earthquake magnitudes measured using the Richter scale depend on the type of seismic wave measured, seismograph used, and distance from the epicenter. The Richter scale isn't very accurate, so scientists came up with a better solution.

What is the moment magnitude scale?

Most scientists now use a scale called the moment magnitude scale to measure earthquake size. This scale is based on the amount of energy released in an earthquake. It measures small and large earthquakes accurately. The movement is calculated using the size of the fault rupture, the amount it slips, and the strength of the broken rocks. ✔

✔ Reading Check

6. Name What is the best scale scientists have for measuring earthquake size?

_____

_____

Academic Vocabulary
structure (STRUK chur)
(noun) a building or something constructed

 **Reading Check**

7. Explain Why does an earthquake lose energy?

_____

_____

Earthquake Intensity

Earthquakes can also be described by the amount of damage they cause. Scientists examine the effects of an earthquake on <u>structures</u>, the land surface, and people's reactions to the shaking. In this way, they can assign intensity values for the earthquake.

Where is an earthquake most intense?

An earthquake becomes more intense the nearer one gets to the earthquake's epicenter. The earthquake loses energy as it spreads out and its waves penetrate rock layers. ☑

Differences in the kind of rocks or sediment affect the amount of shaking. As seismic waves travel, they cause loose sediment to shake more than rocks.

On what do scientists plot intensity values?

Scientists assign intensity values. The values are plotted on a map to give a more complete picture of the earthquake's effects. Data are contoured much like the data on a topographic map. Contour lines join points of equal intensity.

Features of Earthquakes

Seismic waves are measured by scientists to determine where an earthquake occurred and how large it was. Magnitude scales measure the energy released by earthquakes. The intensity of an earthquake is determined by the amount of shaking and damage that occurs in a specific area.

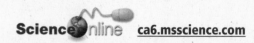 **Science** nline   ca6.msscience.com

Earthquakes

chapter 6

lesson ❹ Earthquake Hazards and Safety

Grade Six Science Content Standard. 2.d. Students know earthquakes, volcanic eruptions, landslides, and floods change human and wildlife habitats. **Also covers:** 1.g.

● Before You Read

Think about the different kinds of damage an earthquake might cause. Write a description on the lines below of the actions and precautions you could take to make yourself safe. Then read the lesson to find out ways to protect against and reduce earthquake damage.

_____

_____

_____

● Read to Learn

Earthquake Hazards

The ground movement during an earthquake doesn't cause many deaths or injuries. The shaking of things humans build does the most damage. Most injuries result from the collapse of buildings or other structures. Fires, landslides, loose sediments, and tsunamis are hazards that often result from earthquakes.

How do earthquakes cause fires?

Fire is the most common hazard to occur after an earthquake. Fires might start when an earthquake ruptures gas lines or causes electrical lines to fall. Broken water pipes can keep firefighters from fighting the fires.

How do earthquakes cause landslides?

When ground is steeply sloped, an earthquake can cause a landslide. A landslide is the sudden movement of soil and rocks down a slope. Landslides can block roads, damage and destroy homes, and damage gas pipes and electrical lines. ☑

MAIN ❮ Idea

Effects of an earthquake depend on its size and the types of structures and geology of the region.

What You'll Learn
- how earthquakes cause damage
- ways to reduce earthquake damage
- ways to make a classroom and home more safe

◄ Mark the Text

Highlight Main Ideas
Highlight any important facts or helpful hints.

☑ Reading Check

1. **Define** What is a landslide?

_____

_____

_____

What is liquefaction?

The process by which shaking makes loose sediment behave like a liquid is called **liquefaction**. When liquefaction occurs under buildings, the buildings can sink into the soil and collapse. ✓

How do earthquakes cause tsunamis?

An earthquake under the ocean can make the seafloor move suddenly. The movement causes powerful waves that can travel for thousands of kilometers. Scientists call ocean waves caused by earthquakes seismic sea waves, or **tsunamis** (soo NAM meez). Far from shore, these waves might be long and flat, and not so dangerous. But when they reach shore, they form a towering crest up to 30 m in height.

What's a warning sign a tsunami is coming?

Before a tsunami hits shore, water might move back rapidly toward the sea, and <u>expose</u> a huge shoreline. This could mean a tsunami will strike soon. You should move to higher ground immediately.

So many earthquakes occur around the Pacific Ocean that tsunamis threaten shorelines constantly. A warning center in Hilo, Hawaii alerts people of an approaching tsunami. The Indian Ocean does not currently have a warning system.

Avoiding Earthquake Hazards

Recall that most earthquakes occur on faults along plate boundaries. The closer you are to one of these faults, the greater the chance of future earthquake damage. Areas where deep, loose sediments are at the surface are also at greater risk for damage than areas with solid rocks at the surface. If you learn where faults are located and loose sediment is at the surface of the Earth, you can avoid these places to avoid damage and injury. ✓

How do scientists find where hazards are greatest?

Seismologists can easily locate fault lines. They also have geologic maps that show where solid rocks and loose sediments exist at the Earth's surface. Scientists use this information to make maps such as the one on the next page. These maps show where earthquake damage might occur. Other maps show areas at risk of landslides, liquefaction, or tsunamis caused by earthquakes.

✔ **Reading Check**

2. Describe a problem caused by liquefaction.

_____

_____

_____

Academic Vocabulary
expose (eks POHS) (verb) to uncover or reveal

✔ **Reading Check**

3. Locate Where are most earthquake hazards likely to occur?

_____

_____

_____

_____

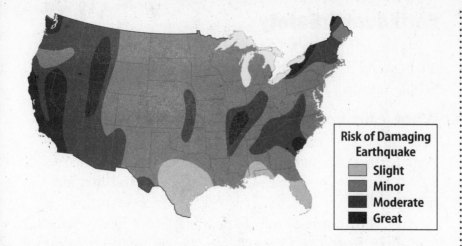

Risk of Damaging Earthquake
- Slight
- Minor
- Moderate
- Great

Picture This

4. Determine Circle the areas of the country where an earthquake plan is needed the most.

How do people plan ahead for earthquakes?

If cities know where earthquakes often occur, they can plan around them when they decide how to use land. Planners may use land at greater seismic risk for agriculture or parks instead of homes.

Earthquakes and Structures

During earthquakes, buildings, bridges, and highways can be damaged or destroyed. Most earthquake deaths occur when these structures crumble.

What types of buildings are safer in an earthquake?

Different types of buildings experience different levels of damage during an earthquake. Buildings made of brittle materials like non-reinforced concrete, brick, or adobe can be severely damaged. Buildings made of flexible material, like wood, suffer less damage in an earthquake. Short buildings withstand earthquakes better than tall buildings.

What types of buildings have engineers created to stand up to earthquakes?

Today in California, some new buildings are supported by flexible, circular anchors placed under the buildings. The anchors are made of alternating rubber and steel plates. The rubber acts as a cushion to absorb earthquake waves. Tests have shown that buildings supported by these anchors should be able to withstand a magnitude-8.3 earthquake without major damage. In older buildings, workers install steel rods in building walls to add support. These measures protect buildings in areas that are likely to experience earthquakes.

FOLDABLES™

D Explain Make a two-tab Foldable. Label the tabs as illustrated. Explain how each of these earthquake preparation methods reduces earthquake damage and loss of life under the tabs. Include information about how each method is put into action in California.

Informed Land Decisions

Earthquake-Resistant Structures

Earthquake Safety

People can plan ahead to protect themselves before, during, and after an earthquake. Planning ahead helps people respond better in a crisis.

What can you do before an earthquake?

To protect yourself and your family, you need a plan. Prepare an earthquake supply kit that contains canned food, water, a battery-powered radio, a flashlight, and first aid supplies. ✔

Make your home as safe as possible. Move heavy objects from high shelves to lower shelves. Learn how to turn off the gas, water, and electricity in your home. To reduce the risk of fire, make sure all gas appliances are held securely in place.

What can you do during an earthquake?

If you are indoors when an earthquake strikes, you should stay put. Move away from windows and any objects that might fall on you. Stay in an interior doorway or under a sturdy table or desk. If you're outdoors, stay in the open. Don't go near power lines or anything that might fall.

What can you do after an earthquake?

Stay calm. If gas lines are damaged and leaking, get an adult to turn off the gas. If you smell gas, leave the building and call for emergency help. Stay away from beaches. Tsunamis can happen after the earth has stopped shaking. ✔

Surviving an Earthquake

Buildings that cannot withstand the shaking of an earthquake cause most of the damage and loss of life. Earthquakes cause fires, landslides and tsunamis.

Scientists cannot predict exactly where and when an earthquake will occur. The best way to protect yourself and your family is to be prepared. Store emergency supplies and learn how to secure appliances. You should also know how to turn off electrical power, gas, and water in your home. Like a fire drill, practicing an earthquake plan could reduce the effects of an earthquake on you and your community.

✔ **Reading Check**

5. Summarize What should your earthquake supply kit contain?

_____

_____

_____

✔ **Reading Check**

6. Explain Why should you stay away from beaches after an earthquake?

_____

_____

Science Online   ca6.msscience.com

Volcanoes

lesson ❶ Volcanoes and Plate Boundaries

Grade Six Science Content Standard. 1.d. Students know that earthquakes are sudden motions along breaks in the crust called faults and that volcanoes and fissures are locations where magma reaches the surface. **Also covers:** 1.e.

● Before You Read

Think about pictures of volcanoes that you have seen. On the lines below, describe what a volcano looks like. Read this lesson to learn how volcanoes are formed.

_____

_____

● Read to Learn

What is a volcano?

A <u>volcano</u> is a land or underwater feature that forms when magma reaches Earth's surface. Recall from Chapter 2 that magma is molten, liquid rock material that is below the surface of Earth. Magma is less dense than the solid rock around it. It is also under a lot of pressure. For these reasons, magma tends to move upward, toward the surface of Earth. Once magma reaches Earth's surface, it is called lava.

How do volcanoes form?

Recall from Chapter 3 that some of Earth's <u>internal</u> heat is left over from when the planet first formed. When rocks become hot enough under pressure they can melt and become magma. In order for a volcano to form, magma must first reach Earth's surface.

The figure on the next page shows magma inside Earth being forced to the surface. Magma is less dense than the rock from which it melted. Therefore magma tends to rise above the denser rock. The density of magma depends upon its composition, the amount of dissolved gas, and the temperature of the magma. Magma also tends to be more buoyant than the rocks around it. The buoyant force pushes magma toward Earth's surface. ☑

MAIN ‹ Idea

Most volcanic activity occurs along plate boundaries.

What You'll Learn
- the cause of volcanic activity
- how volcanoes affect the shape of Earth's surface

◀ Mark the Text

Identify Answers. Read each question heading. Then underline the answer to the question in the text.

Academic Vocabulary
internal (ihn TUR nul) (adj) inside; within

✓ Reading Check

1. **Determine** Which two words best describes magma compared to other rock? (Circle your answer.)
   a. hot and dense
   b. hot and buoyant

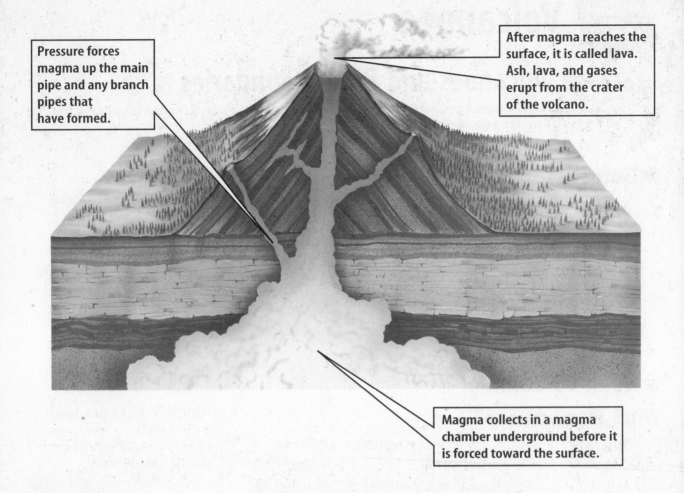

Pressure forces magma up the main pipe and any branch pipes that have formed.

After magma reaches the surface, it is called lava. Ash, lava, and gases erupt from the crater of the volcano.

Magma collects in a magma chamber underground before it is forced toward the surface.

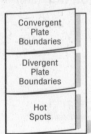

FOLDABLES

Ⓐ Sketch and Record
Make a three-tab Foldable. Label the front tabs, as shown. Underline important words as you read about boundaries and hot spots. Record what you learn under the tabs.

Convergent Plate Boundaries

Divergent Plate Boundaries

Hot Spots

Where do volcanoes occur?

Volcanoes are not common in all regions of the world. They only occur where conditions favor the release of magma from Earth's interior. Conditions that favor the formation of a volcano are usually related to plate boundaries. Recall that plate boundaries are the areas where there is movement of Earth's lithospheric plates. The map at the top of the next page shows the locations of volcanoes, hot spots, and plate boundaries around the world.

What happens at convergent plate boundaries?

Convergent plate boundaries are where two lithospheric plates are being pushed together. When one or both plates are under the ocean, the colder plate sinks beneath the warmer one as shown at the bottom of the next page. Sometimes the magma finds its way to the surface and forms one or more volcanoes.

Volcanoes, Hot Spots, and Plate Boundaries

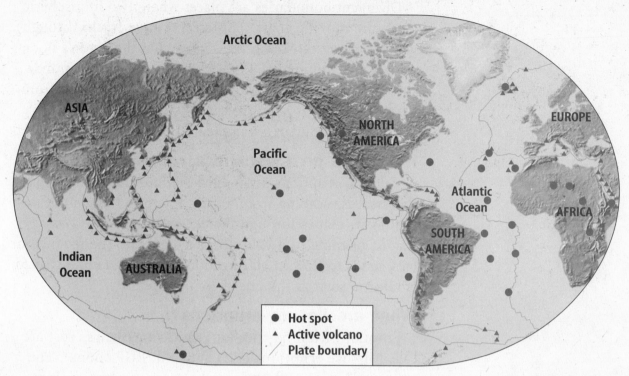

Volcanic Arcs When an oceanic plate sinks beneath a continental plate, a volcanic arc is formed. A volcanic arc is a string of volcanoes that forms on the edge of the continent.

Island Arcs When two oceanic plates are pushed together, a long, curved string of volcanic islands may form. These are called island arcs.

Picture This

2. **Analyze** Highlight plate boundaries on the map. Notice the volcanoes along the way. Now look for hot spots. Describe where they are found.

Magma is less dense than rock, so it is forced upward and eventually erupts from the volcano.

As the oceanic plate slides downward, rock melts and forms magma.

Picture This

3. **Interpret** Circle the area where melted rock is forced upward.

Where do rift valleys form?

Divergent boundaries are places where two lithospheric plates are moving apart. This creates huge cracks through which magma can reach the surface. On the ocean's floor, underwater mountains, known as mid-ocean ridges, may surround these cracks. Mid-ocean ridges are formed from hardened lava. On land, volcanic activity at divergent continental boundaries can create rift valleys.

How does heat escape a volcano?

Volcanic eruptions are one of the most noticeable signs that heat is escaping from Earth's interior. Magma releases heat as it escapes through the central, circular, or oval-shaped opening of a volcano, called a **vent**. The magma, or lava, tends to flow in all directions, creating a cone-shaped landform, such as the one shown above.

What are fissure eruptions?

Eruptions from narrow, long cracks in Earth's crust are called **fissure eruptions** (FIH shur • ih RUP shunz). The magma flows smoothly along the crack resulting in long, sheet-shaped landforms.

When fissure eruptions occur at a divergent plate boundary underwater, they can form mid-ocean ridges and new seafloor. Fissure eruptions can also occur on land, at divergent continental boundaries. Fissure eruptions on land produces new crust at Earth's surface. ☑

✔ **Reading Check**

4. Locate Where on Earth does a fissure eruption create new crust?

_____

_____

What forms away from plate boundaries?

Some volcanoes are not close to any known plate boundaries. The places they are found are known as hot spots. A **hot spot** is where the temperature between Earth's core and mantle is particularly hot. This can melt rock and lead to volcanic activity. The Hawaiian Islands, for example, are located on a hot spot. Scientists continue to study how hot spots move and form.

Formation of Volcanoes

Heat and pressure from Earth's interior cause rock to melt and become magma. Because magma is less dense than the surrounding rock, the buoyant force causes it to rise to Earth's surface. Most volcanoes form at divergent or convergent plate boundaries. However, hot spots form away from plate boundaries. Scientists continue to study hot spots to better understand how and why they form.

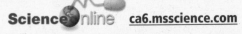

Science Online   ca6.msscience.com

chapter 7 Volcanoes

lesson 2 Volcanic Eruptions and Features

Grade Six Science Content Standard. 1.f. Students know how to explain major features of California geology (including mountains, faults, volcanoes) in terms of plate tectonics.

● Before You Read

Describe what happens when you squeeze a tube of toothpaste. Now imagine squeezing the same tube if it were full of water. Would it flow the same way or differently? Explain on the lines below. Read the lesson to learn about how the composition of magma influences the results of volcanic eruptions.

_____

_____

_____

● Read to Learn

What controls volcanic eruptions?

All lava does not flow the same way. Some types flow quickly and easily, while others are thick and slow moving. The composition of magma controls how lava flows and the way a volcano erupts.

What is the composition of magma?

Scientists can predict the energy of a volcanic eruption based on the percentage of silica and oxygen that is present in the magma. **Viscosity** (vihs KAH suh tee) is a physical property that describes a material's resistance to flow. The viscosity of magma depends on its composition. Silica increases magma's viscosity. In other words, magma that contains a lot of silica will have a high viscosity. Magma with high silica content is thick and sticky. High silica content will cause magma to flow slowly, like honey or frosting.

Magma that contains only a little silica and more iron and magnesium will have low viscosity. Magma with low silica content is thin and runny and flows much more easily, like warm syrup.

MAIN ⟨Idea

The composition of magma determines the different types of lava flow and volcanic features produced.

What You'll Learn
■ the internal processes of a volcano
■ the different types of volcanic landforms
■ about California volcanic activity

◀ **Mark the Text**

Identify Meanings
Highlight the meaning of each underlined term in the lesson.

FOLDABLES™

B Record Information
Make a layered Foldable and use it to record what you learn about erosion and deposition under the tabs. Give examples of each as they relate to your personal experiences.

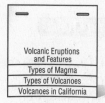

Volcanic Eruptions and Features
Types of Magma
Types of Volcanoes
Volcanoes in California

What other factors affect volcanic eruptions?

The temperature of magma also <u>impacts</u> the way a volcano erupts. In general, magma is less viscous at high temperatures. That means that the higher the temperature of magma, the more easily it flows.

The last important factor that affects a volcanic eruption is the amount of trapped gas that magma contains. Trapped gases can include water vapor, carbon dioxide, sulfur dioxide, and hydrogen sulfide. The more gas magma contains, the more explosive its eruption is likely to be, even if the composition of the magma would suggest that is should have a quiet eruption. ☑

Types of Magma and Lava

Two types of magma are basaltic (buh SAWL tihk) magma and granitic (gra NIH tihk) magma. These two types differ in how much silica they contain.

What is basaltic magma?

Basaltic magma contains a low percentage of silica. It typically has a low viscosity, meaning it flows freely. Basaltic magma is a thinner, more fluid magma. When basaltic lava erupts, it tends to flow quietly and produce quiet eruptions.

Basaltic lava that erupts from a volcano tends to pour from the vent and run down the sides of the volcano. As this *pahoehoe* (pa HOY hoy) lava cools, it develops a smooth skin and forms ropelike patterns. If the same lava flows at a lower temperature, a stiff, slowly moving *aa* (AH ah) lava forms. Basaltic lava that erupts underwater forms bubble-like pillow lava.

What is granitic magma?

Granitic magma contains a high percentage of silica. When granitic magma emerges on Earth's surface, it typically will have high viscosity, meaning it flows slowly. Granitic lava will be sticky and lumpy. It also tends to trap gases, which causes pressure to build up and produce explosive eruptions. ☑

Types of Volcanoes

Three main types of volcanoes are found on Earth's land surface. They are shield volcanoes, cinder cone volcanoes, and composite volcanoes. There are also some volcanoes that do not fit into any of those categories.

✔ Reading Check

1. List What three factors affect the way in which magma erupts?

_____

_____

_____

✔ Reading Check

2. Identify What are the properties of granitic magma? (Circle your answer.)
a. slow flowing
b. thin and ropelike

What are shield volcanoes?

When basaltic lava erupts, it tends to spread out in flat layers. This creates shield volcanoes. **Shield volcanoes** are huge, gently-sloping landforms made from basaltic lava. The Hawaiian Islands are made almost entirely of shield volcanoes. A shield volcano is illustrated below.

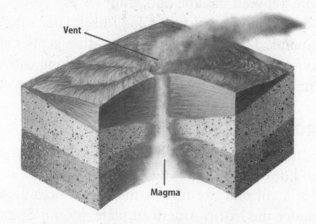

Picture This

3. Determine Use your pen or pencil to draw how lava erupts out of the shield volcano in the figure.

What are cinder cone volcanoes?

Cinder cone volcanoes are made of **tephra** (TEH fruh), which is any solid material erupted from a volcano. Tephra can be as small as ash particles or as large as huge boulders.

How are cinder cones formed? It usually happens during explosive volcanic eruptions. During such eruptions, lava is thrown high into the air. It breaks apart and hardens into tephra. This tephra falls to the ground near the vent, forming a steep landform in the shape of a cone, as shown in the figure below.

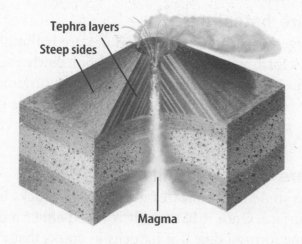

Picture This

4. Illustrate Use your pen or pencil to draw how lava erupts from the cinder cone volcano in the figure.

What are composite volcanoes?

Composite volcanoes are made from layers of both lava and tephra. The layers of a composite volcano build up from the alternation of quiet and explosive volcanic eruptions. The quiet and explosive eruptions occur because the composition of magma in a composite volcano is somewhere between basaltic and granitic. ☑

What do composite volcanoes look like? They often form tall mountains. The top is very steep, with a gentler slope at the bottom. Composite volcanoes form along convergent plate boundaries where one plate sinks beneath the other.

Volcanoes in California

There are many volcanoes in California. Two of the most well known are Mount Shasta and Lassen Peak. They are part of a volcanic arc that formed as an oceanic plate moved under the North American plate. Mount Shasta is a composite volcano. Lassen Peak is a giant lava dome.

Intrusive Volcanic Features

Most magma never reaches Earth's surface. Instead, much of it cools underground. Magma forms rocks known as intrusive igneous rock bodies. These are always formed underground. Over time, however, these rocks may become visible at Earth's surface due to erosion.

There are many types of intrusive igneous rock bodies. The most common types are batholiths, sills, dikes, and volcanic necks.

What are batholiths?

The largest intrusive volcanic features are batholiths. Batholiths form when magma bodies cool slowly deep beneath Earth, without ever reaching Earth's surface. They harden and form enormous rocks. Batholiths can be hundreds of kilometers in width and length. They can be several kilometers thick.

What are dikes and sills?

Magma sometimes squeezes into cracks in rocks below Earth's surface. Once it hardens it is called either a dike or a sill. Magma forms a dike if it hardens in cracks that cut across rock layers. Magma creates a sill if it hardens in cracks that run parallel to rock layers. Some dikes and sills are only a few meters in length. Others are hundreds of meters long. ☑

☑ **Reading Check**

6. Compare What is the difference between a dike and a sill?

_____

_____

_____

Other Volcanic Features

What happens when a volcano stops erupting? Magma hardens inside the vent. This creates a large, solid igneous rock that is surrounded by the rest of the volcano. Eventually the volcano erodes away. The igneous rock that remains is known as a volcanic neck.

What are lava domes?

A very different landform is created when granitic magma erupts. This slow-moving lava piles up and creates a lava dome. Lava domes are rounded volcanic features made when lava is too viscous to move freely. Lava domes often have dangerous, explosive eruptions. This is because gases build up in its thick lava. Lava domes often appear in the central vents of composite volcanoes.

What are lava tubes?

As swiftly moving magma flows from a magma chamber, its outer surface can cool, harden, and form a lava tube. When the magma stops flowing or changes direction, the lava tube drains and becomes a rocky, hollow tube. A lava tube can be over 8 m in diameter and as long as 48 km in length. Some lava tubes form underground.

What are calderas?

Sometimes after an eruption, the top of a volcano can collapse. This produces a large depression called a caldera (cal DAYR uh). Calderas often fill with water and turn into beautiful lakes. ☑

Characteristics of Volcanic Eruptions

The way a volcano erupts is controlled by the composition of the volcano's magma, the amount of dissolved gas in the magma, and the magma's temperature. These factors also affect the viscosity and flow of lava, which result in different volcanic features. Shield volcanoes, cinder cone volcanoes, and composite volcanoes all are common types of volcanoes. Fissure eruptions also release heat from Earth's interior through lava flows. The Cascade Range, located across California, Oregon, and Washington, contains potentially active volcanoes.

Think it Over

7. Predict Is a volcanic neck dangerous? Why or why not?

_____

_____

_____

_____

_____

Reading Check

8. Explain What happens to many calderas?

_____

_____

Volcanoes

chapter 7

lesson ❸ Hazards of Volcanic Eruptions

Grade Six Science Content Standard. 2.d. Students know earthquakes, volcanic eruptions, landslides, and floods change human and wildlife habitats.

MAIN Idea

Volcanic eruptions change habitats.

What You'll Learn

- about geologic events that scientists observe to help predict volcanic eruptions
- types of technology that help scientists monitor volcanoes

Mark the Text

Identify Main Ideas

As you read this lesson, underline the main idea in each paragraph.

Reading Check

1. **Define** What is an active volcano?

_____

_____

_____

● Before You Read

On the lines below, describe a smoggy day. What did it look and feel like? Read the lesson to learn how volcanic eruptions can affect air quality, as smog does, but in a much more devastating way.

_____

_____

_____

● Read to Learn

Effects on Habitats

Smog created by cars and industry can cause air pollution. Volcanic eruptions affect air quality too. However, the effects of volcanic eruptions often last longer and are more destructive than those due to smog. Volcanic eruptions can also harm habitats in a variety of ways.

Are there active volcanoes in the western United States?

There are many active volcanoes in the western United States. A volcano is active if it could erupt at any time. However some volcanoes are more active than others. For example, Mount St. Helens in Washington is an active volcano. It erupted violently in 1980, and has had many minor eruptions since then. Lassen Peak and Mount Shasta are also examples of active volcanoes. Both Lassen Peak and Mount Shasta have erupted in the last 10,000 years. When a volcano remains quiet for a long time, people tend to forget that it may erupt again. However eruptions and environmental hazards can still occur. ☑

What is volcanic ash?

During the 1980 Mount Saint Helens eruption, a large amount of <u>volcanic ash</u> was released into the atmosphere. Volcanic ash is large amounts of fine-grained tephra made of tiny mineral and glasslike particles. Volcanic ash is abrasive. When mixed with water it becomes slippery and heavy.

Volcanic ash can severely damage property. When layers of ash build up on rooftops, the weight can cause structural damage. Volcanic ash can also fall onto wildlife habitats. Ash can bury plants and animals and their food sources and contaminate the water supply.

How does volcanic ash cause landslides and mudflows?

When volcanic ash mixes with water it can become a heavy mix of materials that flows quickly downhill. This can happen with heavy rains, or when ash mixes with melting water from glaciers or snow. The flow of material is known as a <u>lahar</u> (LAH har). Some of the largest lahars begin as landslides. Landslides can occur from volcanic eruptions, earthquakes, precipitation, or gravity. Volcanic ash, tephra, dirt, rock, and even trees can mix with ground water and precipitation and form a lahar. Rivers of debris can move downhill at rates up to tens of meters per second. It is not possible for humans to outrun a swiftly moving lahar.

When human habitats are built in river valleys near volcanoes, the flow of debris is directed toward the town. A fast-moving lahar provides little time for warning. The effects can be disastrous. Both human and animal habitats can be buried beneath these heavy mudflows.

How do the gases from volcanic eruptions impact living things?

Volcanic eruptions can release gases that are dangerous to living things. For example, sulfur dioxide and hydrogen sulfide mix with water to form sulfuric acid precipitation. This is harmful to both plants and animals.

What are pyroclastic flows?

In some cases, hot gases mix with volcanic ash and solids, forming a <u>pyroclastic flow</u>. A pyroclastic flow moves quickly down the sides of a volcano and continues downhill. Pyroclastic flows tend to follow valleys and can destroy everything in their paths. <u>Intensely</u> hot gases that travel within the pyroclastic flow can also contaminate the air.

FOLDABLES™

C Record Make a Foldable table. Record what you learn about the causes and effects of each environmental hazard.

	Volcanic Ash	Landslides and Lahars	Gases	Pyroclastic Flows	Lava Flows
Cause					
Effect					

💡 Think it Over

2. Explain Why can a lahar be dangerous?

_____

_____

Academic Vocabulary
intense (ihn TENTS) (adj) great energy

What are the characteristics of lava flows?

Lava flows can destroy human and wildlife habitats by starting fires, destroying property and crops, and releasing smoke, which affects air quality. Lava flows cannot be controlled, but most lava flows move slowly enough that humans can be warned to leave the area.

Predicting Volcanic Eruptions

Most volcanoes show signs of activity before they erupt. Warning signs can include small earthquakes, emission of gases, and changes in the shape or temperature of the ground. Scientists monitor volcanoes in an attempt to predict dangerous eruptions. ☑

Small Earthquakes Usually before a volcanic eruption, there are many small earthquakes ranging from about 1 to 3 in magnitude. To predict earthquake activity, scientists build networks of ground-based seismic detectors.

Gas Emissions Different amounts of gases—carbon dioxide, for example—indicate how deep below the surface magma is located. Changes in the type or amount of gases coming from a volcano might signal that an eruption will occur soon.

Ground Movement and Temperature As magma moves toward Earth's surface, the ground around the volcano can bulge and increase in temperature. Using remote-sensing devices, scientists are able to detect changes in temperature and ground movement

Monitoring Volcanic Activity

Using satellites in space, scientists can detect even small changes in the shape and temperature of the land surface around a volcano. Satellites are also used after volcanic eruptions, to track the movement of clouds of volcanic ash. This saves lives and avoids costly damage to equipment. ☑

Volcanic Hazards

Volcanic landforms and features range in shape and size. All types of volcanoes can emit gases, solids, lava, and tephra during explosive eruptions. Gas emissions, lahars, and pyroclastic flows are some hazards that result from volcanic eruptions. Through technology, scientists are better able to monitor volcanic activity from space. This helps them more accurately predict dangerous eruptions.

☑ **Reading Check**

3. List What are three signs that a volcano might soon erupt?

_____

_____

_____

_____

☑ **Reading Check**

4. Explain How do satellites help scientists predict volcanic eruptions?

_____

_____

_____

_____

Science Online **ca6.msscience.com**

Weathering and Erosion

lesson ❶ Weathering

Grade Six Science Content Standard. 2.a. Students know water running downhill is the dominant process in shaping the landscape, including California's landscape.

● Before You Read

Have you ever taken a walk and kicked pieces of rock as you walked? Did you wonder where the rocks came from and why there were so many different sizes and shapes of rocks? On the lines below, describe what you think caused the soil and rocks to form on Earth's surface. Then read the lesson to learn about the processes of weathering.

_____

_____

MAIN ‹Idea

Rocks exposed at Earth's surface are broken down into sediment and soils by weathering.

What You'll Learn
■ the difference between chemical and physical weathering
■ the effects of weathering
■ how humans and living things affect weathering

● Read to Learn

What is weathering?

<u>Weathering</u> is the destructive process that slowly breaks down and changes rocks exposed at Earth's surface. Weathering is caused by the action of water, wind, ice, and gravity. Water, wind, ice, and gravity are referred to as agents of weathering. These agents create two processes that change rocks. These processes are chemical weathering and physical weathering.

Chemical Weathering

<u>Chemical weathering</u> is when minerals and rocks at Earth's surface break down from exposure to water and gases in the atmosphere. This exposure causes the composition of the minerals of a rock to change. The result is the formation of new minerals. Think about old cars that are rusted. Gases in the air such as carbon dioxide, oxygen, and water vapor combine with the metal of a car to form rust. These gases also dissolve minerals and rocks. These are examples of chemical weathering.

◀ Study Coach

Ask Questions As you read, jot down questions that you would like to ask the author if you had a chance. Pose these questions to the class or to your teacher as a basis for discussion.

FOLDABLES

Ⓐ Compare and Contrast Make a Venn-diagram Foldable and use it to compare and contrast chemical and physical weathering under the appropriate tabs. Describe what they have in common under the center tab.

What is the most common agent of chemical weathering?

The most common agent of chemical weathering is water. When rocks and minerals dissolve in water, they are soluble. Carbonic acid is formed when water is mixed with carbon dioxide from the air. This is the same weak acid found in a carbonated soft drink. Rainwater is slightly acidic because it contains dissolved carbon dioxide from the air. When water contains carbonic acid it is much more effective at weathering than if the water does not contain carbonic acid. ☑

How does acid contribute to chemical weathering?

Can you guess what will happen when slightly acidic rainwater comes in contact with rock? The rainwater reacts with the minerals in rock such as feldspar. When feldspar weathers rapidly, the feldspar changes into clay minerals. The formation of clay is one of the most common results of chemical weathering.

Human-made pollution can cause chemical weathering to occur even more rapidly. For example, when coal is burned, sulfur dioxide is <u>released</u> into the atmosphere. Sulfur dioxide is turned into sulfuric acid when it combines with water vapor in the air. This mixture eventually becomes acid rain. Acid rain damages rocks, building, plants, soils, and lakes.

How does oxygen contribute to chemical weathering?

Oxidation (ahk sih DAY shun) occurs when oxygen that is dissolved in water comes in contact with metals, such as iron. Oxidation of iron is called rust. Rocks that have iron in them are sometimes oxidized. Iron rocks are rusty red in color. ☑

How do different rock types react to chemical weathering?

The type of rock also affects how quickly its surface is chemically weathered. For example, think about old headstones in a cemetery. The details carved into some headstones are still clear even after 100 years. Those detailed headstones are made of rock that resists chemical weathering, such as granite or slate. Some headstones have lost most of their original carved details after 100 years. Those headstones are probably made of rock that is affected by chemical weathering, such as limestone or marble.

✔ **Reading Check**

1. **Explain** What is the effect of carbonic acid in water?

_____

_____

Academic Vocabulary
release (rih LEES) (verb)
to let go or set free

✔ **Reading Check**

2. **Define** What is oxidation?

_____

_____

_____

_____

Physical Weathering

Physical weathering is the breaking of rock into smaller pieces without changing its mineral composition. Processes of physical weathering include frost wedging and the work of plants and animals.

What is frost wedging?

Frost wedging is when water freezes, expands, and melts in the cracks of rocks. As you might know, water expands when it freezes. So when water seeps into cracks of rocks and the temperature drops below freezing, the water will then freeze and expand. This expansion causes pressure to build and forces the crack to open slightly, as shown in the figure below. If there are many freezing-thawing cycles and the crack is wedged completely open, the rock will break into pieces.

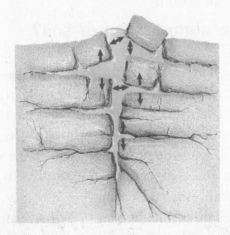

How do plants and animals contribute to physical weathering?

Physical weathering can also be caused by plants and animals. A sidewalk or driveway that is broken or buckled near a large tree is an example of how a plant can cause <u>physical</u> weathering. When the tree grows, the roots get larger. Eventually, the roots become so large that they push up on the concrete and cause it to crack. Plant roots in search of water also can grow into cracks within rocks. As plant roots grow in size, they eventually wedge the rocks apart.

Animals can cause physical weathering as well. When animals burrow, they move loose rocks and dirt to the surface. This material is then exposed to wind and water, and more weathering occurs.

💡 Think it Over

3. Name one example of physical weathering and one example of chemical weathering.

_____

_____

_____

_____

Picture This

4. Determine Use a marker to highlight the place in the figure where ice wedging is most likely to force the surrounding rock to fall away.

Academic Vocabulary

physical (FIH zih kul) (adj.) experienced through the senses and the rules of nature

Soil Formation

Rocks that have been broken and weathered produce a layer of soil or dirt. <u>Soil</u> is a mixture of weathered rock, minerals, and organic matter. The organic matter in soil is made up of decaying plants and animals. Soil is formed in several ways. The formation of soil is affected by several factors. The type of rock, the climate, the length of time a rock has been weathering, and the actions of plants and animals all contribute to soil formation. ☑

Recall that feldspar is the most common mineral in rocks. Feldspar often breaks down through chemical weathering and forms clay minerals. That is why clay is one of the most common ingredients in soils.

Plants need the nutrients in soil in order to grow. In addition to being anchored in soil, when plants grow in soil, they keep it from eroding.

What is the composition of soil?

Residual soil is what soil is called when it remains in the same place where it is formed. The composition of the soil will match the composition of the rock from which it formed. For example, granite contains quartz. Quartz is resistant to weathering. A soil that develops from granite will be sandy because of the sand-sized grains of quartz it contains. But a soil developed from basalt, which contains large amounts of feldspar, will have sticky clay particles instead. When soil is transported by wind, water or glaciers to a new location, the composition of the soil does not match the composition of the rock beneath it. ☑

What are soil layers?

Have you ever dug a deep hole in the ground? If so, you might have noticed layers that look different from each other. These layers are called soil horizons. Soil horizons take thousands of years to develop. Together, the soil horizons make up a soil profile. An illustration of the soil horizons in a forest is shown on the next page.

A horizon The topmost soil layer is called the A horizon. It contains humus, which is small rock and mineral particles, along with decomposed plant material. The layer is usually a dark color. Water flowing through this horizon removes minerals from it. When this happens, it can cause the A horizon to become light colored because it has been stripped, or leached, of minerals.

✔ Reading Check

5. **Describe** What is soil?

_____

_____

_____

✔ Reading Check

6. **Define** What is residual soil?

_____

_____

_____

B horizon The minerals that have been removed, or leached, from the soil are deposited in the next layer, called the B horizon. Because this layer contains large amounts of clay, it is red or brown.

C horizon The soil layer called the C horizon is below the B horizon. Partly weathered parent material or bedrock makes up the C horizon. Below the C horizon of soil is unweathered parent material, solid rock.

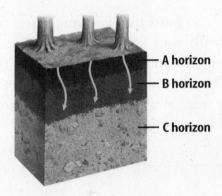

— A horizon
— B horizon
— C horizon

Picture This

7. Determine Trace over the arrows that show the direction of leaching, or stripping the soil of its minerals.

Weathering and Landforms

The processes of chemical weathering and physical weathering work together to break rocks into smaller pieces. Chemical weathering changes the mineral and chemical composition of rocks. Physical weathering breaks down rocks without changing the composition. Rocks that have been broken into smaller pieces by physical weathering have more surface area that can be exposed to chemical weathering. This process helps to form soils. Soils develop in layers called horizons. Weathering plus other factors influence the character of the soil that forms.

Weathering and Erosion

lesson ❷ Erosion and Deposition

Grade Six Science Content Standard. 2.b. Students know rivers and streams are dynamic systems that erode, transport sediment, change course, and flood their banks in natural and recurring patterns. **Also covers:** 2.a, 2.c.

MAIN Idea

Movement of rock and soil are natural occurrences caused by specific geologic conditions.

What You'll Learn

- how the land surface is changed by the action of water
- how streams are formed and reshape the land
- how mass movements relate to land use in California

Study Coach

Make Flash Cards Write each underlined word on one side of a flash card. Write the definition on the other side. Use the cards to guide yourself.

FOLDABLES

Ⓑ Record Information
Make a layered Foldable and use it to record what you learn about erosion and deposition under the tabs.

| Erosion and Deposition |
| Mass Wasting |
| Rivers |
| Ocean Waves |
| Glaciers |
| Wind Deposits |

● Before You Read

Sometimes it rains so hard that you might feel the soil is disappearing beneath your feet or that your house might slip and slide on its foundation. How sturdy is the Earth's surface? On the lines below, describe what you think causes Earth's surface to wear away. Then read the lesson to learn about erosion and deposition.

_____

_____

_____

● Read to Learn

What are erosion and deposition?

Flowing water from a river or a stream can move pieces of rock and soil particles downstream. This process of moving material from one location to another is called **erosion**. Erosion can be caused by running water, rain, waves, glaciers, or wind. Gravity causes landslides to erode land. **Deposition** is the process of sediments being laid down in a new location.

Mass Wasting

Erosion caused by gravity is call **mass wasting**. Mass wasting occurs when the ground becomes soaked with water. Then rocks and soil begins to move downhill because the soil is too weakened to hold them in place. Mass wasting can also be started by earthquakes or other events that shake the earth. The rumble of heavy machinery, blasting, or thunder can begin a mass-wasting event. The steeper the slope or hill, the more likely mass wasting will occur.

What is fast mass wasting?

<u>Landslides</u> occur when rapid events move soil, loose rocks, and boulders. Mudslides are a form of landslide. A mudslide is a mixture of soaked soil and rock material. Rock falls are when loosened rock falls from steep cliffs. Slumps, such as the one shown below, occur when a block of rock and the overlying soil slide down a slope as one large mass.

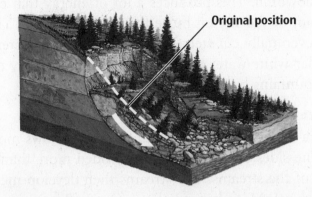
Original position

What is creep?

Sometimes mass wasting takes a long time to occur. Creep is the slowest type of mass wasting and occurs when sediment moves slowly down a hillside. You might see signs of creep along the highway. Look for creep in the tilting of telephone poles, trees, or fences in a downward direction.

How does climate affect erosion?

The amount of water an area receives is determined by its climate. Regions that receive large amounts of rainfall are more likely to have mass wasting than areas with dry climates. Climate also affects vegetation. Thick vegetation, such as plants and trees, often prevents landslides because the root systems of plants help to hold dirt and rock in place. This is especially helpful on slopes where a heavy rainfall will cause water to run downhill quickly, taking rocks and soil with it. This cushion of vegetation softens the falling raindrops. Water then gently soaks into the soil, reducing erosion. ☑

Water and Erosion

What happens to water that does not soak into the ground? It flows over Earth's surface into lakes, streams, and rivers. The surface water eventually ends up in the oceans. Streams and rivers erode the land by transporting sediment and depositing sediment into new locations.

Picture This

1. **Interpret** What force could not have caused the slumped material to move downslope? (Circle the answer.)
   a. erosion
   b. water
   c. gravity
   d. deposition

☑ **Reading Check**

2. **Explain** Why do areas with thick vegetation have less erosion than areas with little vegetation?

_____

_____

_____

What are the stages of stream development?

As you know, waterfalls and river rapids occur in steep mountain regions, not in flat valleys. This is because river systems change as water moves from high in the mountains down to lakes or oceans at sea level.

Streams are formed in mountains by rainfall and melting snow. The steep slopes of the mountainside cause the water to rush downhill. This produces a lot of energy that erodes the bottom of the stream more than the sides of it. These streams eventually cut steep, V-shaped valleys and create spectacular white water rapids and waterfalls as they move down mountains onto the land below.

How do meanders develop?

When a stream flows over gentle slopes, it flows more slowly. The sides of the streams are eroded more than the bottoms of the streams. The streams then develop **meanders** (me AN durs). Meanders are the curves in the stream. Because the speed of the water is greatest at the outside of the bend, the meanders tend to become wider and wider with passing time. On the other hand, water flows more slowly on the inside of the meanders and sediment is dropped there. This process is called deposition.

How does water affect deposition?

Recall that deposition is when sediment and rock are eroded and transported by river systems. Eventually this material is deposited in a new location, such as further downstream. If the water comes upon a shallower slope, it slows down. Once water is slowed down, it reduces the amount of energy that it takes for the stream to carry sediment.

What features develop from deposition?

Deposited sediments can form distinct features. For example, if a river deposits material on the inside of a meander, it can cut off a large U-shaped meander from the river. This action creates a small lake called an oxbow lake. Once a stream or river reaches a larger body of water, such as a lake, it slows down. Then the sediment drops out and forms a triangular-shaped deposit called a delta. ☑

Sometimes rivers empty from steep, narrow canyons onto flat plains at the foot of mountains. When that happens, a similar triangular deposit is formed. This feature is called an alluvial fan.

Think it Over

3. Compare When you walk slowly, often going here or there instead of in a straight line, you are said to be *meandering*. How is your meandering like meanders that are formed by a stream?

_____

_____

_____

Reading Check

4. Determine How is a delta formed?

_____

_____

_____

_____

_____

What causes flooding?

A <u>flood</u> occurs when the water level in a river rises above normal heights and overflows the sides of the river's banks. Rainstorms or rapid melting of snow can cause floods. Water can then spill onto the <u>floodplain</u>, a wide, flat valley that is <u>located</u> along the sides of some rivers and streams. Floodplains are formed from the side-to-side erosion that occurs from a meandering stream.

Sometimes floods can be helpful to farmers. The sediment-filled floodwaters supply farms with fertile, rich soil that helps crops grow.

How is flooding prevented?

Long, low ridges are formed on the floodplain along both sides of a river. These ridges, called natural levees, are formed when sediment is carried and deposited by floodwaters. These levees protect the area from extreme flooding. Sometimes artificial levees are built along the banks of rivers to help control floodwaters. In New Orleans, a break in its artificial levee allowed the river to flood the nearby region. In urban areas, such as Los Angeles, some streams and rivers are lined with concrete to reduce flooding. ☑

Because floods are unpredictable, building on floodplains or too near dams and levees is not a good idea. All geological factors need to be considered before any construction is begun.

Shorelines and Erosion

California has 1,100 miles of shoreline along the Pacific Ocean. The waves that pound California's shores are incredibly powerful. This constant pounding of waves changes and erodes the shape of the shoreline. This erosion occurs because of rocks breaking into smaller pieces. Pounding waves also transport and grind sediment in the surf zone. The waves then deposit the material farther along the shore.

How are beaches eroded?

A <u>beach</u> is a landform consisting of loose sand and gravel. A beach is located along a shoreline. Most of California's steep shores have been formed by beach erosion. Sand is also supplied by the continuous flow of rivers to the oceans. Sediment carried by the rivers gets deposited on the beach. Wave action then moves the sediment along the shore. ☑

Academic Vocabulary
locate (LOH kayt) (verb)
to find; to identify a particular spot, place

☑ **Reading Check**

5. Explain What is the purpose of an artificial levee?

_____

_____

_____

☑ **Reading Check**

6. State Where does sand on a beach come from?

_____

_____

7. **Identify** Which of the following is not caused by wave erosion? (Circle your answer.)
   a. cliffs
   b. marine platforms
   c. sea arches
   d. sea caves

Picture This

8. **Locate** Circle the arrows that show the direction sediment moves along a shoreline.

What are erosion features?

Shoreline cliffs are caused by erosion. The cutting action of waves at the base of rocks creates cliffs. When a cliff is eroded, it moves back from the shoreline. A flat area, called a wave-cut platform, is left behind. If the platform is lifted above the water level by upward movement along faults, it is known as a marine terrace. Unusual shapes such as sea caves, sea stacks, and sea arches can be formed by erosion when the waves erode the softer rocks. ✔

What is a longshore current?

Waves come to the shore in three steps. First, waves come in at an angle to the shoreline. Second, the friction of striking the beach at an angle causes the waves to end until they are almost parallel to the coast. Finally, waves retreat from the beach perpendicular to the shoreline. This process is called longshore transport. The movement of the water is a longshore current, as illustrated below. A longshore current can move large amounts of sediment along coasts.

Shoreline

Sediment transport

How might erosion be prevented?

One way to limit the erosion of sediment from beaches is through structures called groins. Groins trap sediment that might be moved along the coast by longshore transport.

What are glaciers?

__Glaciers__ are large masses of ice and snow. It takes hundreds to thousands of years for a glacier to form. Glaciers form in areas where the amount of winter snowfall is greater than the amount of summer melting. Glaciers move very, very slowly, at a rate of about 2.5 cm per day.

What are two types of glaciers?

There are two types of glaciers. Valley glaciers, sometimes called alpine glaciers, form in existing stream valleys high in the mountains. Valley glaciers flow from high to low elevations. There are more than 100,000 valley glaciers on Earth today. ☑

Continental glaciers, called ice sheets, are different from valley glaciers. Continental glaciers are several kilometers thick and cover entire land areas. The only continental glaciers on Earth today are in Antarctica and Greenland. Continental glaciers were the types of glaciers that existed during past ice ages.

How do glaciers erode land?

Glaciers erode land by pulling along rocks and boulders that are trapped at the base of the ice. When glaciers pass over Earth's surface, the rocks pulled along by the glacier create grooves and scratches. This is similar to the way sandpaper leaves scratches on wood. Scientists examine these grooves left on Earth's surface by glaciers to determine the direction the glacier once moved. ☑

How is sediment deposited by glaciers?

As glaciers melt, they deposit sediment that had been frozen in the ice. The sediments deposited by glaciers are called till and outwash. Till often builds up along the sides and fronts of glaciers into long, high ridges called moraines. Till can form interesting landforms because it might have been molded beneath the glacier. Outwash consists mostly of sand and gravel. Sometimes these materials are quarried for use in construction.

Wind

Have you read about the Great Dust Bowl of the 1930s? This event occurred when a long drought devastated the southern Great Plains of the United States for an entire decade. People during that time had plowed their fields deeply and allowed pastures to become overgrazed. These poor agricultural practices left the soil unprotected and exposed to the elements. Strong winds removed this soil. Records from that time report that skies were blackened by great wind-generated dust storms.

✔ Reading Check

9. Locate Where do valley glaciers form?

_____

_____

_____

✔ Reading Check

10. Describe How do glaciers erode land?

_____

_____

_____

_____

How do wind, erosion, and deposition interact?

Wind lifts and redeposits loose material. There are two common types of wind-blown deposits. One type of wind-blown deposit is a sand dune, such as the one shown below. Sand dunes are formed from mounds and ridges created when heavier sediment blows along the ground surface. Eventually the sediment gets pushed into piles and dunes form. ☑

Loess (LUHS) is the second type of wind-blown deposit. Loess consists of wind-blown silt that was carried in the air. Loess is the smallest grain size produced by glacial erosion. Strong winds that blow across glacial outwash pick up the loess and redeposit it elsewhere. As wind-blown sediment is carried along, it cuts and polishes exposed rock surfaces.

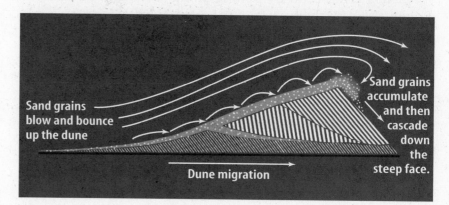

Sand grains blow and bounce up the dune

Sand grains accumulate and then cascade down the steep face.

Dune migration

Shaping by Erosion and Deposition

Several geologic processes are involved in erosion and deposition. Mass wasting causes landslides, rock falls, and mudslides. Climate and the amount of rainfall an area receives are directly related to mass wasting. Rivers erode streambeds and transport sediment to new locations. The great power of this erosion changes the shape of their streambeds. Wave action on ocean shores breaks up rocks and creates distinct features along beaches. Erosion and deposition by glaciers create familiar mountain scenery. Wind can also be strong enough to cause erosion and to form dunes. You can see the results of all these processes in the landscapes of California today.

Picture This

12. Label Draw an arrow to show the direction the wind is blowing to form this dune.

Weathering and Erosion

lesson ❸ Reshaping the California Landscape

Grade Six Science Content Standard. 2.b. Students know rivers and streams are dynamic systems that erode, transport sediment, change course, and flood their banks in natural and recurring patterns. **Also covers:** 1.f.

● Before You Read

On the lines below, describe your favorite type of landscape in California, such as its mountains, deserts, valleys, or beaches. Then read the lesson to learn more about the geology of California's landscapes.

_____

_____

_____

_____

MAIN ‹ Idea

The geology of California includes mountains, deserts, valleys, and coastlines.

What You'll Learn

■ how California's landscape compares to other places

■ how weather, weathering, and erosion affect California's landscape

■ how California's land use is unique

● Read to Learn

Mountain Landscapes

California can be divided into four types of landscapes: mountains, deserts, the Central Valley, and the coast. Mountain ranges cover most of California. Many of the rocks that make up mountain ranges formed below Earth's surface hundreds of millions of years ago. Tectonic processes have uplifted and exposed the rocks at the surface.

What are glaciated mountains?

During the last 2.5 million years, glaciers have carved many features in the Sierra Nevada and Klamath Mountains, making them into glaciated mountains. Glaciated mountains have features caused by glacial erosion such as U-shaped valleys and hanging valleys. Those features are common in Yosemite National Park.

Recall that glaciers deposit material as well as erode and move rock. Moraines are features that were produced when the glaciers melted and deposited their till on glaciated mountains.

Mark the Text

Highlight Key Terms
Highlight key terms and their meanings as you read this lesson.

FOLDABLES™

● Record Information
Make a four-tab Foldable and label the front tabs as illustrated. Record key concepts you learn about California's mountains, coats, deserts, and central valley.

California Mountains

California Coasts

California Deserts

California Central Valley

What are other erosional features of mountains?

California mountain ranges have other types of erosional features, too. Some of these erosional features are not related to glaciers. As you have learned, streams and rivers change as water moves from the mountains to the oceans. Steep white-water streams in V-shaped valleys form in the high, steep parts of mountains. Wider and more meandering rivers are common in the plains. Landslides and rock falls are also common in California's mountains. ☑

Desert Landscapes

California's deserts are primarily located in the southeastern corner of the state. The deserts can be found between many small mountain ranges and extinct volcanoes. These deserts consist of flat, sandy valleys and dry lakebeds called playas. Alluvial fans are common features in the deserts created from deposits. Erosion also occurs in deserts.

What are examples of desert landscapes?

The Mojave Desert has very little vegetation. It is sometimes referred to as the high desert because of its high average elevation. In contrast is the Colorado Desert, which lies as much as 75 m below sea level. The Colorado is referred to as the low desert. Because the Colorado River irrigates the desert, it has become an agricultural area. The Colorado River forms an enormous delta where it empties into the Gulf of California. ☑

What are some features of desert landscapes?

Wind-blown sand dunes are common in the desert. Strong, consistent winds power hundreds of windmills for nearby towns. Windmills are used to generate electricity. The popular vacation community of Palm Springs, California, uses power from windmills.

What is the Basin and Range?

The **Basin and Range** is a large area of north-south trending mountain ranges and valleys. The Basin and Range is located primarily in Nevada and Utah. Most of this area has a desert climate. At the western edge of the Basin and Range in California is Death Valley.

Death Valley Gold-seekers named Death Valley in 1849 because of the valley's harsh conditions. The part of Death Valley known as Badwater is the lowest point in the western hemisphere. Badwater is 86 m below sea level.

☑ **Reading Check**

1. **Identify** three types of erosional features common in California mountain ranges.

_____

_____

_____

☑ **Reading Check**

2. **Compare** the Mojave Desert to the Colorado Desert.

_____

_____

_____

Features of Death Valley Death Valley is made up of many arroyos. <u>Arroyos</u> are streambeds that only contain water during heavy rains or floods. Because Death Valley has very little vegetation to stop erosion, during heavy rains or floods a great amount of rock and sediment is transported downstream in the arroyos. The deposition of rock and sediment means that Death Valley has many alluvial fans. Some of Death Valley's ancient alluvial fans were formed about six million years ago. In Golden Canyon in Death Valley, ancient alluvial fan deposits have turned into rock. ☑

The Central Valley

The Central Valley of California is about 800 km long and 50 km wide. It is also called the Great Valley. The Central Valley is a fault-bounded valley. This means that the mountains around it have been uplifted along faults, and the Central Valley has dropped to lower elevations.

What are rivers in the Central Valley?

Two <u>major</u> rivers flow through the Central Valley. These rivers have meanders and flow slowly along shallow slopes. The south-flowing Sacramento River is in the north part of the valley. The north-flowing San Joaquin River is in the south part of the valley. These two rivers meet and form a delta into the Pacific Ocean through San Francisco Bay.

What effect does deposition have on the Central Valley?

Rivers flow into the Central Valley and deposit abundant sediment from the surrounding mountains. This sediment provides the valley with a thick fertile soil. In fact, the Central Valley is one of the most productive agricultural areas in California. The Central Valley provides half the produce in the United States.

Coastal Landscapes

California is known for its sandy beaches, but these beaches can change significantly during various seasons. For example, the beach at Point Reyes National Seashore changed in just six months. Stormy El Niño conditions during the 1997–1998 winter caused many landslides and erosion near the shore at Point Reyes. Waves then transported and deposited more sediment along the shore.

✓ Reading Check

3. Identify Which feature is the result of Death Valley's lack of vegetation? (Circle the answer.)
   a. arroyos
   b. alluvial fans

Academic Vocabulary
major (MAY jur) (adj.) of greater importance; prominent

💡 Think it Over

4. Determine In what way do rivers help the Central Valley provide half the produce in the United States?

_____

_____

_____

_____

_____

California's Landscapes

California has a variety of landscapes. Mountains are formed either by tectonic uplift or by the formation of volcanoes. Glaciers, streams, wind, and mass wasting have carved the mountains into the shapes you see today. Deserts experience strong winds and erosion that form dunes. Tectonic activity formed the Basin and Range. Fertile soils and wide rivers cover the Central Valley. The coast of California has beaches and rocky shorelines. ☑

✔ **Reading Check**

5. Summarize Which forces carved California's mountains?

_____

_____

_____

Earth's Atmosphere

lesson ❶ Energy from the Sun

Grade Six Science Content Standard. 4.a. Students know the Sun is the major source of energy for phenomena on earth's surface; it powers winds, ocean currents, and the water cycle. **Also covers:** 4.b.

● Before You Read

Write a few lines about what pleasures we would miss if the Sun did not provide warmth and energy. Then read the lesson to find out how the Sun is Earth's major energy source.

_____

_____

_____

_____

● Read to Learn

Earth's Atmosphere

The **atmosphere** (AT muh sfihr) is a mixture of gases that surrounds the Earth. The atmosphere is made up of several layers. Each layer is different and has different properties. In general, this mixture of gases in the atmosphere is called *air*.

What is the composition of air?

The main chemicals that make up Earth's atmosphere are nitrogen and oxygen. Oxygen gas (O_2) represents 21 percent of the composition of the atmosphere. Humans and other animals need to breathe oxygen to live. Nitrogen gas (N_2) represents 78 percent of the composition of the atmosphere.

Particles and gases such as water vapor (H_2O), argon (Ar), carbon dioxide (CO_2), and ozone (O_3) make up about 1 percent of the atmosphere. Although these substances are present in very small amounts, they are still important. Some of these substances affect weather and climate and protect living things from harmful solar radiation. Others can have damaging effects on the atmosphere and the organisms that breathe those substances in the air.

MAIN Idea

The Sun is the major source of energy for Earth.

What You'll Learn
- differences between layers of the atmosphere
- how solar radiation impacts the Earth

Mark the Text

Identify the Main Point
Highlight the main point of each paragraph. Use a different color to highlight a detail or example that helps explain the main point.

FOLDABLES

A Describe Make a two-tab Foldable. Label the tabs as illustrated. Describe the composition and function of Earth's atmosphere under the tabs.

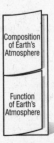

Composition of Earth's Atmosphere

Function of Earth's Atmosphere

What are the layers in the atmosphere?

The atmosphere is made up of layers. The figure below shows the approximate height above sea level where the atmospheric layers are found. The lowest layer of the atmosphere is the troposphere. The **troposphere** (TRO puh sfihr) is the region of the atmosphere that extends from Earth's surface to a height of about 8 km to 15 km. The troposphere is the region of the atmosphere where life on Earth exists. The troposphere has weather and climate. The troposphere also holds the majority of Earth's air. As you go higher in the troposphere the air temperature gets colder. ☑

The stratosphere is above the troposphere. The **stratosphere** (STRA tuh sfihr) is the region of the atmosphere that extends from about 15 km to 50 km. As you go higher in the stratosphere the air temperature gets hotter. This occurs because the concentration of ozone is much higher in the stratosphere than in the troposphere. The layer of ozone in the stratosphere absorbs some of the Sun's harmful ultraviolet radiation, causing air temperature to rise.

The top two layers of the atmosphere are the mesosphere and the thermosphere. The mesosphere extends to about 80 km above Earth's surface. The thermosphere continues to extend far above Earth's surface. It does not have a defined upper limit. Beyond the thermosphere is space.

1. Identify Name the layer of the Earth where living things exist.

Picture This

2. Interpret Look at the graph of atmospheric temperatures. Does the temperature in the thermosphere increase or decrease with altitude?

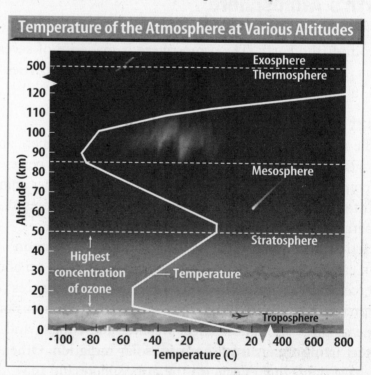

Temperature of the Atmosphere at Various Altitudes

The Sun's Continuous Spectrum

There are many forms of radiation. For example, X-rays, microwaves, radio waves, and visible light are all forms of radiation. All of these forms of radiation can be compared by using the electromagnetic spectrum. The **electromagnetic spectrum** (ih lek troh mag NEH tik • SPEK trum) includes the entire range of wavelengths or frequencies of electromagnetic radiation. The electromagnetic spectrum is used to describe differences in radiation, from long waves to short waves. Ninety-nine percent of solar radiation <u>consists</u> of ultraviolet light, visible light, and infrared radiation.

What is visible radiation?

Sometimes we refer to sunlight as visible light or white light. Wavelengths in the visible range are those you can see. Have you ever used a prism to separate white light into different colors? White light can be divided into the colors red, orange, yellow, green, blue, indigo, and violet. Visible light, which includes all of the colors of a rainbow, is actually visible radiation. The energy coming from the Sun peaks in the range of visible light. ☑

What is near-visible radiation?

In addition to visible light, we also receive infrared and ultraviolet radiation from the Sun. The wavelengths of these two forms of radiation are just beyond the range of visibility to human eyes. However, these forms of radiation can still be detected by some organisms.

<u>Infrared (IR) waves</u> have longer wavelengths than visible light and sometimes are felt as heat. If you have ever felt the warmth from a fire, you have felt infrared radiation. You also can feel infrared radiation when you are being warmed by the Sun as you lie on the beach. Some snakes, such as rattlesnakes, have special sensors near their eyes that can detect infrared radiation. The snakes use this ability to detect the heat given off by warm-blooded prey at night. ☑

<u>Ultraviolet</u> (ul truh VI uh luht) (UV) <u>waves</u> have shorter wavelengths than visible light. Humans do not see or feel ultraviolet radiation. However, you might have felt the effects of ultraviolet radiation. Ultraviolet light is the radiation that is responsible for causing skin to tan or sunburn. UV radiation reaches Earth's surface on both sunny and cloudy days. Some animals such as bees, butterflies, and birds can detect ultraviolet light with their eyes.

Academic Vocabulary
consist (kahn SIST) (verb) to be composed or made up of

☑ **Reading Check**

3. **Determine** How are visible light and visible radiation related?

_____

_____

_____

☑ **Reading Check**

4. **Identify** Which kind of radiation is sometimes felt as heat by humans?

_____

_____

How does sunlight penetrate the atmosphere?

As the Sun's radiation passes through the atmosphere, some of it is absorbed by gases and particles and some of it is reflected back into space. As a result, not all the radiation coming from the Sun reaches Earth's surface. Study the figure below. You can see that about twenty percent of incoming solar radiation is absorbed by gases and particles in the atmosphere. Oxygen, ozone, and water vapor all absorb incoming ultraviolet radiation. Ozone in the stratosphere absorbs much of the incoming ultraviolet radiation from the Sun. Some of the infrared radiation from the Sun is absorbed by water and carbon dioxide in the troposphere. However, the wavelengths of visible light are not greatly absorbed by Earth's atmosphere.

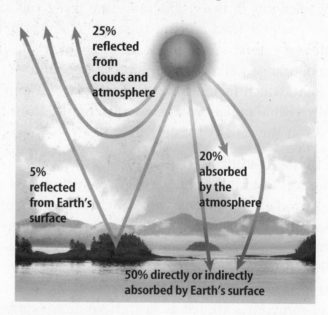

25% reflected from clouds and atmosphere

5% reflected from Earth's surface

20% absorbed by the atmosphere

50% directly or indirectly absorbed by Earth's surface

The Sun's Power

The Sun emits an enormous amount of radiation. The radiation from the Sun has to travel a very long distance through space before reaching Earth. Although this is true, solar heating is responsible for the climate conditions on Earth. The conditions at Earth's surface are suitable for life as we know it because of the atmosphere and the Sun's radiation.

Why is the Sun a constant source of energy?

The Sun will continue to produce energy for billions of years. That is why we consider the Sun a constant source of energy. Most of the time solar radiation reaching Earth from the Sun is nearly uniform. ☑

✔ Reading Check

6. Explain why energy from the Sun is considered a constant source of energy.

_____

_____

_____

What is the effect of the angle of sunlight?

Although the radiation and heat leaving the Sun is constant and uniform, some places on Earth get more of the heat and energy. A beam of sunlight near the equator is almost perpendicular to Earth's surface. The beam of sunlight is concentrated into a small area. Since the sunlight is in a small area, the land, water, and air become warm.

The same size beam of sunlight also strikes Earth's surface near the poles. But because sunlight is distributed over Earth's curved surface, it strikes at a low angle. The beam of sunlight is spread out over a larger area. The land, water, and air near the poles do not warm as much. When the Sun stays below the horizon during the winter months, the poles can become very cold.

What is solar energy on Earth?

The Sun's energy is responsible for climate and weather and makes it possible for people to live on Earth. For example, air currents, or wind, are generated as the Sun's energy heats the air. Wind leads to the formation of waves on the surface of the ocean. Powerful weather systems including hurricanes and tornadoes get their energy from the Sun.

The energy that drives the water cycle comes from the Sun. The water cycle is the cycle in which water at Earth's surface continually evaporates and returns to Earth's surface as precipitation. The Sun is also necessary for photosynthesis. Photosynthesis produces plant materials that release energy when broken down.

Humans use energy from the Sun in many ways. Solar energy can be collected with solar collection devices that capture the rays coming from the Sun. The water cycle is powered by the Sun. The energy from fast-flowing rivers that are part of the water cycle can be transformed into electrical power with the use of dams. The Sun's energy creates wind that can power windmills. The windmills convert energy from wind into electrical energy. ☑

The Sun's Energy

Constant energy from the Sun reaches Earth in the form of visible light, infrared radiation, and ultraviolet light. Solar radiation warms water, air, and land at Earth's surface. It powers the water cycle and photosynthesis in living organisms that form the base of the food chain.

Think it Over

7. **Summarize** why some parts of the Earth are warmer than others.

_____

_____

_____

_____

Reading Check

8. **Identify** two ways that people use the energy from the sun to make electrical power.

_____

_____

_____

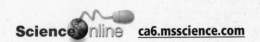
Science Online   ca6.msscience.com

Earth's Atmosphere

lesson 2 Energy Transfer in the Atmosphere

Grade Six Science Content Standard. 4.d. Students know convection currents distribute heat in the atmosphere and oceans. **Also covers:** 3.c, 3.d.

MAIN Idea

Earth's atmosphere distributes thermal energy.

What You'll Learn

- why hot air rises and cold air sinks
- the properties of radiation
- the effects of greenhouse gases

Mark the Text

Underline Definitions
Underline the definitions of all the underlined terms.

Picture This

1. **Describe** the movement of the air.

_____

_____

_____

_____

_____

● Before You Read

On the lines below, write a sentence describing what happens to the air above a pot of boiling water. Then read the lesson to find out more about how heat energy is transferred in Earth's atmosphere.

_____

_____

● Read to Learn

Conduction in Air

There are three types of heat transfer—conduction, convection, and radiation. Conduction takes place when molecules transfer their kinetic energy, or thermal energy, by collisions with other molecules.

How is air heated from below?

Conduction is the process that heats air close to Earth's surface. The radiant energy from the Sun warms up the land and the oceans, as shown below. Air is a poor heat conductor. As a result, the hot ground only warms a thin layer of air above it.

Warm air
Cool air

Convection in Air

Convection is the transfer of kinetic energy, or heat energy, through the movement of molecules from one part of a material to another. For example, the hot air that is close to Earth's surface moves to higher altitudes by convection. As a result, air in the troposphere moves all the time. In the troposphere, the movement of air by convection is mostly in a vertical direction.

What happens when air is heated?

As the temperature of air increases, the kinetic energy of the molecules increases. The air molecules are moving faster and becoming farther apart. When this happens, the air expands. When the air expands its density decreases.

How do hot and cold air interact?

When air that is close to the surface warms, it rises. The same process is used to lift a hot-air balloon from the ground. Why does a balloon need hot air to rise? The balloon rises because the air inside the balloon is hotter and, therefore, less dense than the air surrounding it. The balloon will keep rising as long as colder air with higher density remains beneath it. When the temperature inside the balloon is equal to the temperature outside the balloon, it will stop rising.

What happens if the air inside the balloon is colder than the air surrounding it? In this case, the air inside the balloon is denser than the air around it. The denser air moves down while air with lower density remains above it. The balloon with colder air in it sinks.

What are air circulation patterns?

As you have just learned through the balloon example, hot air rises and cold air sinks. The density of the air inside the balloon compared to the air surrounding the balloon causes the balloon to rise or fall. This same principle also works when heating a room. Convection currents created by the heated air circulate throughout the room. Warm air near a heater on the floor rises as its density decreases. The colder air near the ceiling flows down to replace the warmer air that is rising. When the colder air moves closer to the heater, it warms up, so eventually it will move up again. When air rises, it cools down, and then sinks again. This leads to a continuous pattern of circulating air. ☑

B Draw Make a two-tab Foldable. Label the tabs as illustrated. Sketch a convection current and describe what happens as air is heated and cooled under the tabs.

Air is cooled.

Air is heated.

Reading Check

2. Identify What causes air circulation patterns?

_____

_____

Academic Vocabulary
distribute (dihs TRIH byewt)
(verb) to scatter, spread out

What is a convection current?

You have learned that hot air moves up and cold air moves down. This creates a continuous movement of air. The continuous vertical movement of air that occurs in a circular pattern is called convection current. Convection currents <u>distribute</u> heat energy within the troposphere.

What happens when an inversion occurs?

In most cases, the higher you go in the troposphere, the lower the temperature. But in some cases, the warmer air is above the cooler air. The normal air pattern is turned upside down. When something is turned upside down, it is inverted. An **inversion** takes place when warm air sits on top of cold air. When an inversion occurs, the air rising from Earth's surface can only reach a certain altitude until it becomes trapped by the warm layer of air above it. ☑

An inversion can have serious effects. Air might contain a harmful substance such as a pollutant. It would be best for the air to rise as high as possible so that humans and other animals would not have to breathe the polluted air. When there is an inversion, the air cannot move up and away from people. Harmful substances remain trapped close to Earth's surface.

Inversions can happen many times a year. Location, weather conditions, and other factors affect where and when inversions take place. The Los Angeles area often experiences inversions that trap pollutants near Earth's surface. The pollutants in the air reduce visibility.

✔ Reading Check

3. Describe what happens in a temperature inversion.

_____

_____

_____

_____

Radiation Traveling Through Space

Radiation is the transfer of energy in the form of electromagnetic waves. Unlike convection and conduction, radiation can travel through empty space. Solar radiation in the form of electromagnetic waves can travel through space and reach Earth. ☑

At one time, people started fires by rapidly spinning a wooden stick. Friction heated the vibrating wood pieces. Eventually, a fire started. Today, people have learned that the Sun's energy can provide heat. If you place a magnifying glass in a way to capture the sunlight, the area below the magnifying glass will be heated. The magnifying glass focuses the Sun's rays on a small area. The air molecules under the magnifying glass vibrate rapidly as they are warmed. Eventually, the air will become very hot.

✔ Reading Check

4. Explain How does solar radiation get to Earth?

_____

_____

_____

Which types of radiation do different materials absorb?

Different molecules absorb different types of radiation. For example, oxygen and ozone absorb ultraviolet radiation. Water and carbon dioxide absorb infrared radiation.

About seven percent of the Sun's radiation is ultraviolet radiation. Ultraviolet radiation can harm animals and plants. Much of the damage from ultraviolet radiation is absorbed by gas molecules in the atmosphere. Most harmful ultraviolet radiation does not reach Earth's surface.

Do all materials emit radiation?

The Sun is not the only object in the universe to emit radiation. All objects in the universe that have temperatures above absolute zero emit radiation. For example, your own body emits infrared radiation—known as heat.

Earth emits radiation in a similar way as the Sun. The main difference is in the wavelengths of the radiation emitted. Most of the radiation emitted by the Sun is in the visible range. Most of the radiation emitted by Earth is in the infrared portion of the electromagnetic spectrum. ☑

What maintains the radiation balance on Earth?

Earth does not get hotter and hotter from the heat it receives from the Sun. There is a balance between the radiation received from the Sun and radiation leaving Earth. The total amount of energy reaching Earth from the Sun is equal to the amount of energy leaving Earth. Solar radiation that reaches Earth's surface is absorbed by the land and oceans. After the solar radiation is absorbed, it is given off by Earth, mostly as infrared radiation.

What are greenhouse gases?

Gases in the atmosphere absorb some of the infrared radiation given off by Earth. These gas molecules give off the radiation in all directions. Some of the radiation that is given off is directed back to Earth. A greenhouse works in a similar way. The glass of the greenhouse traps the radiation and the greenhouse stays warm.

When gases in Earth's atmosphere direct radiation back toward Earth's surface, this produces an additional warming of Earth's atmosphere. Gases that strongly absorb a part of Earth's outgoing radiation are called **greenhouse gases**. Some greenhouse gases are water vapor, methane gas (CH_4), and carbon dioxide. ☑

✔ Reading Check

5. **State** the difference between radiation from Earth and from the Sun.

_____

_____

_____

_____

✔ Reading Check

6. **Explain** how greenhouse gases warm the atmosphere.

_____

_____

_____

Academic Vocabulary
data (DAY tuh) (noun) factual information used to reason or answer questions

Picture This

7. Identify Circle the arrows that show the heat trapped near Earth's surface by greenhouse gases in the atmosphere.

What is global warming?

Some scientists have gathered <u>data</u> that shows Earth's average surface temperature is increasing every year—a condition called **global warming**. One possible explanation for this average temperature increase is the large amount of greenhouse gases being released into the atmosphere as shown below. Carbon dioxide is given off when gas, oil, and coal are burned. Increases in the amount of greenhouse gases could lead to the additional heating of Earth's surface.

Energy in the Troposphere

Energy from the Sun moves through the troposphere by conduction, convection, and radiation. Air close to Earth's surface is heated by conduction. Convection currents are caused by differences in the density of air masses. These currents circulate vertically and distribute thermal energy within the troposphere. Radiation from the Sun heats Earth's surface unevenly, with more energy concentrated near the equator and less energy near the north pole and south pole.

Radiation received from the Sun is balanced against radiation that Earth gives off. This balance helps keep Earth's surface temperature stable. Greenhouse gases trap some of the radiation given off by Earth. These gases affect Earth's surface temperatures. Some scientists hypothesize that as the amount of greenhouse gases being released into the atmosphere grows, Earth's average surface temperature could increase. This increase in average temperature is referred to as global warming. ☑

Reading Check

8. Explain What do greenhouse gases affect?

_____

_____

_____

Science Online   ca6.msscience.com

Earth's Atmosphere

lesson ❸ Air Currents

Grade Six Science Content Standard. 4.d. Students know convection currents distribute heat in the atmosphere and oceans. **Also covers:** 4.a.

● Before You Read

Write about a time you played in the wind or saw wind damage after a storm. Then read the lesson to learn more about how energy from the Sun creates air currents.

_____

_____

_____

● Read to Learn

Local Winds and Eddies

The Sun is the major source of energy that powers winds. **Wind** is air that is in motion relative to Earth's surface. A current of water or air that runs counter to the main current— especially a circular current—is called an eddy. Sometimes you can see eddies. They might appear as whirlwinds or dust devils when they carry leaves or dust through the air.

What causes uneven heating of Earth's surface?

The uneven heating of Earth's surface causes differences in air pressure. These differences in air pressure over different parts of Earth's surface bring about winds.

The materials found at Earth's surface absorb or reflect different amounts of sunlight. Materials such as soil, rock, and sand absorb solar radiation and become warm in bright sunlight. Other materials, like snow, grass, or trees often remain cool even when the Sun is shining brightly on them. They reflect much of the solar radiation. Because materials absorb or reflect sunlight, Earth's surface is heated unevenly. The air directly above Earth's surface is also heated unevenly. ✔

MAIN ⟨Idea

Solar energy is responsible for air currents and thermal energy.

What You'll Learn

- how solar energy gives rise to winds
- why Earth's surface is heated unevenly
- how pressure differences affect winds
- how air currents circle Earth

Study Coach

Summarize As you read this lesson, stop after each paragraph and summarize the main idea in your own words.

FOLDABLES™

ⓒ Record Information
Make a three-tab concept map. Label the tabs as illustrated and use it to record three key points about wind. Take notes and record new terms under the tabs.

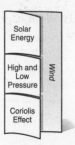

What are updrafts?

Sometimes a large area of land absorbs more solar radiation than the land nearby. When the land becomes warm, it heats the air above it. As the air is heated, it <u>expands</u> and becomes less dense than the surrounding air. Eventually, the heated air rises as part of a convection current. A rising column of air is called an **updraft**.

Updrafts are also called thermals. Many birds use thermals to soar high in the air. They spread their wings to catch the rising air. Birds can remain in the rising column of air for long periods of time without flapping their wings. People also use thermals for hang gliding and parasailing. When air rises in an updraft or thermal, the air at Earth's surface briefly decreases in density and pressure. The surrounding air will move in to fill areas of low pressure.

What are downdrafts?

A **downdraft** is a sinking column of air that occurs when dense air sinks toward Earth's surface. A downdraft that is rapid and forceful is called a downburst.

The dense air that moves down from a downdraft or downburst temporarily creates an area of high pressure at Earth's surface. This occurs because the dense air is concentrated on a small area. Eventually, the dense air will move away from the area of high pressure. ☑

How is air affected by pressure?

When air moves rapidly from an area of high pressure to an area of low pressure, we call it wind. As warm air rises, the air pressure close to Earth's surface decreases. The opposite happens when cold air sinks—air pressure close to Earth's surface increases. Therefore, wind moves away from areas of high pressure and toward areas of low pressure.

Air Currents Around Earth

As the air above Earth's surface is heated, it becomes less dense than colder air. Air pressure is lower where air is heated. This generates wind as air moves from areas of high pressure to areas of low pressure. Winds transfer heat and water vapor from one location to another which, in turn, affects weather and climate.

There is more to learn about wind. Not all winds are explained by dense air sinking at the same time as less-dense air is rising. Earth's rotation also affects winds. ☑

Academic Vocabulary
expand (ihk SPAND) (verb)
to increase the extent, number, volume, or scope of

1. Describe What causes a downdraft or downburst?

_____

_____

_____

2. Explain What is one of the causes of wind?

_____

_____

_____

What is the Coriolis effect?

Moving air and water appears to turn to the right in the northern hemisphere and to the left in the southern hemisphere, as shown in the figure below. This change in the movement of air and water, known as the **Coriolis** (kor ee OH lus) **effect**, is caused by Earth's rotation.

The differences in the amounts of solar radiation received on Earth's surface and the Coriolis effect combine to create wind patterns on Earth's surface. These global winds move heat and water vapor around Earth's surface and affect weather.

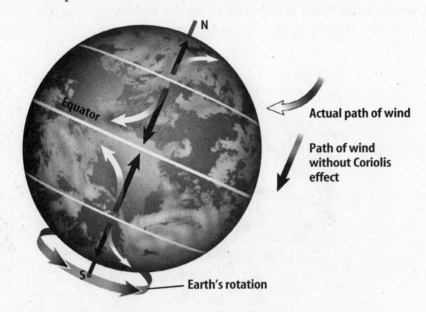

Actual path of wind

Path of wind without Coriolis effect

Earth's rotation

Picture This

3. Identify Trace the arrows north of the equator that shows the path of wind after the Coriolis effect.

How do global convection currents occur?

Scientists use a three-cell model to describe the circulation of Earth's atmosphere. The three cells in each hemisphere look like donuts wrapped around Earth.

The first cell is a convection cell called the Hadley Cell. Hot air rises at the equator and moves to the top of the troposphere. Then, the air moves toward the poles until it cools and sinks back to Earth's surface near 30° latitude. The first cell is completed when most of the air returns toward the equator near Earth's surface.

The third cell, or polar cell, is also a convection cell. Cold, dense air from the poles moves toward the equator along Earth's surface. The air becomes warmer and eventually rises near 60° latitude. The second cell, or Ferrel cell, between 30° and 60° latitude, is not a typical convection cell. Its motion is partially driven by the other two cells, like rolling cookie dough between your two hands. These three cells exist in the southern hemisphere as well. ☑

Reading Check

4. Summarize Which convection cells exist in the southern hemisphere?

_____

_____

_____

What is the effect of global convection cells?

The three global convection cells in each hemisphere create northerly and southerly winds. When the Coriolis effect acts on the winds, they turn and blow to the east or the west, creating relatively steady, predictable winds. Sailors have used the trade winds and the westerlies for centuries to sail across the ocean. Sometimes sailors found little or no wind near the equator. That is because, as the Sun heats the air and water near the equator, the air rises, creating low pressure and little wind. The area near the equator at which there is little wind is referred to as the doldrums. Earth's winds are shown in the figure below.

Picture This

5. Identify Which winds are located on either side of the Equatorial doldrums?

_____

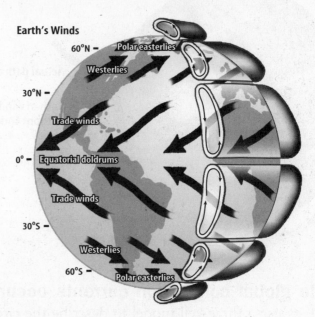

Earth's Winds

60°N — Polar easterlies
Westerlies
30°N —
Trade winds
0° — Equatorial doldrums
Trade winds
30°S —
Westerlies
60°S — Polar easterlies

What are jet streams?

<u>Jet streams</u> are strong, continuous winds that move at speeds of 200 km/h to 250 km/h at the top of the troposphere. There is a polar jet stream and a subtropical jet stream in each hemisphere. Jet streams affect weather. They also affect speeds of airplanes and flight patterns.

Air Currents at Earth's Surface

Earth's surface and the air above it are heated unevenly by the Sun. This creates pressure differences and causes wind. The Coriolis effect influences the direction in which wind flows. <u>Distinct</u> wind patterns can be found around the globe. Three circulation cells exist in each hemisphere—Hadley cells, polar cells, and Ferrel cells. Wind moves heat and water vapor throughout the atmosphere.

Think it Over

6. Determine A 200 km/h wind over the Arctic would most likely be what kind of wind? (Circle your choice.)
a. a polar jet stream
b. a subtropical jet stream
c. the Coriolis effect

Academic Vocabulary
distinct (dih STINKT) (adj.)
separate; notable

Science Online   ca6.msscience.com

Oceans

lesson ❶ Earth's Oceans

Grade Six Science Content Standard. 7.f. Read a topographic map and a geologic map for evidence provided on the maps and construct and interpret a simple scale map.

● Before You Read

Imagine you are a great blue whale swimming in the ocean and diving deep below the surface. What do you think you might see? Write your ideas on the lines below. Read the lesson to learn how scientists map the ocean.

_____

_____

_____

_____

● Read to Learn

Mapping Earth's Oceans

Earth contains five major oceans—the Pacific Ocean, the Atlantic Ocean, the Indian Ocean, the Arctic Ocean, and the Southern Ocean. The Pacific Ocean is the largest ocean. Subduction zones surround the Pacific Ocean in many places. The subduction zone is where the edges of tectonic plates slip past each other. As a result, the Pacific Ocean is slowly decreasing in size. The Atlantic Ocean is smaller than the Pacific. New ocean floor is <u>created</u> in the middle of the Atlantic Ocean by lava continually rising from deep in Earth. The Atlantic Ocean is slowly growing larger.

The Indian Ocean is the shallowest ocean. The Arctic Ocean is at the most northern part of Earth. It is extremely cold in this region and much of the Arctic Ocean is often covered in ice. The Southern Ocean surrounds the continent of Antarctica and extends north to a latitude of 60°S. It connects the Pacific, Indian, and Atlantic Oceans. A map of the world's oceans is shown on the top of the next page.

MAIN ‹ Idea

Mapping the ocean floor helps in understanding Earth's global features.

What You'll Learn

- the different oceans on Earth
- how bathymetric maps of the oceans are made
- the features of the ocean floor

Study Coach

Think-Pair-Share Work with a partner. As you read the text, discuss what you already know about the topic and what you learn from the text.

Academic Vocabulary
create (kree AYT) (verb) to bring into existence

Earth's Oceans

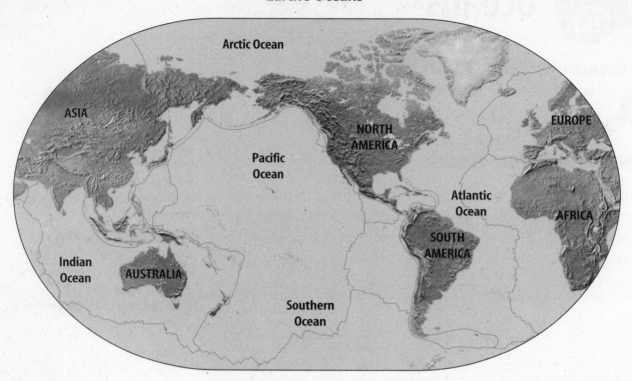

Arctic Ocean
ASIA
NORTH AMERICA
EUROPE
Pacific Ocean
Atlantic Ocean
AFRICA
SOUTH AMERICA
Indian Ocean
AUSTRALIA
Southern Ocean

Picture This

1. Analyze Which ocean connects the Indian Ocean, the Pacific Ocean, and the Atlantic Ocean?

_____

_____

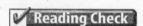 **Reading Check**

2. Identify What is an important reason for mapping the ocean floors?

_____

_____

_____

What are bathymetric maps?

Hidden beneath the water, there are the same kinds of geological shapes that we see on land. Underwater mountain ranges, flat areas, and trenches are on each ocean floor.

If you were sailing around the world you would want to have a good map of the oceans. A good map would tell the depth of water and where dangerous obstacles are under the water. The depth of water is measured from sea level to the ocean floor. **Sea level** is the level of the sea's surface halfway between high and low tides. The **ocean floor** is Earth's surface underneath the ocean water. ☑

Before modern technology, sailors would drop a rope from their ship until it hit the bottom of the ocean. Then they would measure the length of rope they let out and record the water depth. This method of measuring water depth is called sounding. By making a large number of soundings and compiling them, a map of the ocean floor, or a bathymetric (BATH ih meh trihk) map, can be created. A **bathymetric map** is a map of the bottom of the ocean showing the contours of the ocean floor and its geologic features.

What is echo sounding?

Today oceanographers map the ocean floor using sound and radio waves. Sonar **echo sounding** is a sounding that is made using sound waves. Scientists attach an instrument that emits a sound wave to the bottom of a ship. They then measure the time it takes for the sound wave to bounce off the ocean floor and return to the ship. If the sound bounces back quickly, the depth of the ocean is shallow. If the returning sound takes a long time, the depth of the ocean is deep. ☑

Sound waves, radio waves, and light waves can be used to map the locations of coastlines, the geological features on the bottom of the oceans, and the location and direction of currents. Satellites use radio waves to detect small bumps and dips in the ocean surface. These bumps and dips reflect the locations of mountains and trenches on the ocean floor.

The Ocean Floor

Imagine taking a slice through the ocean floor and looking at it from the side. This is called a bathymetric profile, or a cross-section of the ocean. Typical geologic features you will see in a bathymetric profile of the ocean floor are the continental shelf, the continental slope and rise, mid-ocean ridges, trenches, and abyssal plains. Notice these features below.

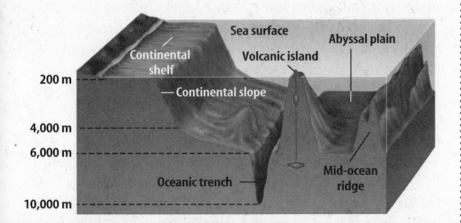

What is the continental shelf?

The **continental shelf** is an underwater portion of continental crust that extends from the edge of a continent and gently slopes toward the deeper parts of the ocean. Along the east coast of the United States, the continental shelf is wide. California has a narrow continental shelf.

3. Explain How does echo sounding work?

_____

_____

_____

_____

_____

_____

Picture This

4. Identify the feature on the ocean floor that appears as small hills or mountains.

_____

_____

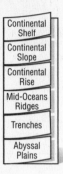

FOLDABLES

A Define and Explain
Make a six-tab Foldable. Label the tabs as illustrated. Define the terms and explain each as they relate to the Earth's ocean under the tabs.

Continental Shelf
Continental Slope
Continental Rise
Mid-Oceans Ridges
Trenches
Abyssal Plains

Reading Check

5. **Explain** What caused the mid-ocean ridges?

_____

_____

_____

What is the continental slope?

The continental slope is the steep slope between the continent and the deep ocean. The continental slope contains deep canyons. Lots of sediments flow down these canyons, sometimes in huge avalanches. These sediments form deposits between the continental slope and the ocean floor known as the continental rise.

Where is the abyssal plain?

Beyond the continental slope and rise, the ocean floor is extremely flat. This region is called the abyssal (uh BIH sul) plain. The abyssal plain is thought to be giant blocks of rock covered with thick layers of sediment.

What are deep ocean trenches?

Deep ocean trenches are extremely deep valleys that extend along the edges of the oceans. The deepest point in the ocean is 11,033 m in the Mariana Trench in the Pacific Ocean. Ocean trenches are subduction zones, places where the tectonic plates are recycled into Earth's interior. Underwater earthquakes can be common in the regions around trenches.

What are mid-ocean ridges?

The main features of the ocean floor are the mid-ocean ridges. The mid-ocean ridges are a continuous chain of underwater volcanoes more than 65,000 km long that extend through all the ocean basins. The mid-ocean ridges rise 2 km above the ocean floor on average. Mid-ocean ridges are places where tectonic plates are moving away from each other and new sea floor is being created. ☑

Features of the Ocean Floor

The five oceans—the Pacific Ocean, the Atlantic Ocean, the Indian Ocean, the Arctic Ocean, and the Southern Ocean—cover more than 70 percent of Earth's surface. The ocean floor has contours and features similar to those found on land, including mid-ocean ridges, trenches, and flat abyssal plains.

The ocean floor can be mapped through echo sounding, using sound waves bounced off the bottom of the ocean. The ocean floor can also be mapped using satellites and radio waves bounced off the surface of the ocean.

Science Online ca6.msscience.com

Oceans

lesson 2 Ocean Currents

Grade Six Science Content Standard. 4.a. Students know the sun is the major source of energy for phenomena on Earth's surface; it powers winds, ocean currents, and the water cycle. **Also covers:** 4.d.

● Before You Read

Think about a time you went swimming or played in a tub of water. You made currents of water when you pushed it around. Write a sentence about how the water moved when you played with it. Read the lesson to learn more about the importance of ocean currents.

_____

_____

_____

● Read to Learn

Influences on Ocean Currents

Ocean water moves from place to place in **ocean currents**, which are like rivers in the ocean. Currents transport water, heat, nutrients, animals and plants, and even ships, from place to place in the oceans.

How do the oceans collect large amounts of energy?

One of the properties of water is that a large amount of heat can be added or <u>removed</u> from it before it changes temperature. The oceans, because they hold huge amounts of water, hold an enormous amount of heat.

It takes a lot of heat to change the temperature of water just a few degrees. In fact, it takes five times more heat to change the temperature of an area of water than it does to change the same area of land. If you have been to the beach on a sunny day, you know that the sand can seem hot, while the water remains cold. Sand changes temperature much more quickly than water does. The oceans, because they are a huge reservoir of water, hold an enormous amount of heat.

MAIN ‹ Idea

Ocean currents help distribute heat around Earth.

What You'll Learn
- how ocean currents are formed
- the global ocean currents and gyres

Study Coach

Create Two-Column Notes As you read, organize your notes in two columns. In the left-hand column, write the main idea of each paragraph. Next to it, in the right-hand column, write details about it.

Academic Vocabulary
remove (rih MOOV) (verb) to take away or change the location of

---

B **Explain** Make a three-tab Foldable. Label the front tabs as illustrated. Describe surface currents, deep ocean currents, and gyres in your own words under each tab, and explain the importance of each.

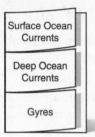

Surface Ocean Currents

Deep Ocean Currents

Gyres

How is heat balanced by the oceans?

Earth receives different amounts of energy from the Sun depending on latitude. The oceans gain the most heat in areas between 30°N and 30°S latitudes. Oceans lose heat at latitudes of 40°.

Water's ability to absorb and lose large amounts of heat energy without changing temperature makes it perfect for moving heat around the planet. In general, ocean currents carry heat from the tropics to the poles. This helps balance the amount of heat across the planet.

What are surface currents?

As the wind blows over the ocean, it tugs on the surface of the ocean, moving the ocean surface water. On windy days, the wind moves the surface water faster than the wave is moving, causing it to crash in front of the wave. This produces whitecaps.

Wind is the most important force moving surface water in the ocean. Wind has the strongest effect on the location and movement of the global ocean currents, shown in the map below.

What does the Coriolis effect do to currents?

Recall that the Coriolis (kor ee OH lihs) effect is caused by Earth spinning on its axis. As a result, winds in the northern hemisphere are deflected to the right and winds in the southern hemisphere are deflected to the left. The spinning of Earth affects liquids in the same way it affects the gases that make up the air. In the northern hemisphere, ocean currents are deflected to the right and in the southern hemisphere, they are deflected to the left. The <u>overall</u> effect is that currents tend to move in a clockwise pattern in the northern hemisphere and in a counter-clockwise pattern in the southern hemisphere.

How does water density affect deep currents?

Some currents are found deep in the ocean where there is no effect from the wind. The density of water produces currents in deep ocean waters. ☑

The amount of salt that is dissolved in a quantity of water is called **salinity** (say LIH nuh tee). As the salinity of water increases, its density increases. The density of water depends on both its temperature and its salinity. Areas of water in different parts of the ocean have different densities. These differences form deep ocean currents. When water on the surface becomes denser than the water below it, the water on the surface sinks.

What causes gyres, the great ocean currents?

A cycle of ocean currents is called a **gyre** (JI ur) There are five major gyres in Earth's Oceans. The North Atlantic and North Pacific Gyres rotate in a clockwise direction. The South Atlantic, South Pacific, and Indian Ocean Gyres rotate in a counterclockwise direction. ☑

Wind is the major influence on the movement of surface water in the ocean. When the North Pacific Current reaches North America, the land mass, as well as the Coriolis effect, pushes the current to the right. The North Pacific Current becomes the California Current, flowing south along the coast of California. When the California Current reaches the tropics, the trade winds move it westward. Its name then changes to the North Equatorial Current. When the North Equatorial Current reaches Asia, the land mass and the Coriolis effect again turn it to the right, and it becomes the Kuroshio Current, moving northward past Japan. When the Kuroshio Current reaches the westerlies, it is pushed toward the east into the North Pacific Current again.

Academic Vocabulary
overall (OH vur awl) (adv.) all over; from one end to the other

✔ **Reading Check**

2. **State** What produces currents in deep ocean waters?

_____

_____

✔ **Reading Check**

3. **Identify** How many gyres exist in Earth's oceans?

_____

What are the special currents and their effects?

The strongest and deepest currents are found on the western sides of the gyres. These currents are called western boundary currents because they are on the western side of the ocean basins. Eastern boundary currents are on the eastern side of the ocean basins.

The biggest of these western boundary currents is the Gulf Stream, which is part of the North Atlantic Gyre. It transports enough water to fill the entire Rose Bowl Stadium about 25 times per second. This water rushes north from the tropics toward the poles. ☑

The Gulf Stream and all the other western boundary currents are important to the redistribution of heat throughout the oceans.

Surrounding the continent of Antarctica is the Antarctic Circumpolar Current, shown in the map below. It is a continuous flow of water, but it is not a gyre because it surrounds land rather than water. It is the largest current in the oceans, with twice as much flow as the Gulf Stream. The Antarctic Circumpolar Current is driven in an eastward direction around the southern part of Earth by the strong westerlies. The Antarctic Circumpolar current serves as a connection between the Pacific, Atlantic, and Indian Oceans.

4. List What are two major special currents that move very large amounts of water in the ocean?

_____

_____

_____

Picture This

5. Name two deep currents shown on the map.

_____

_____

_____

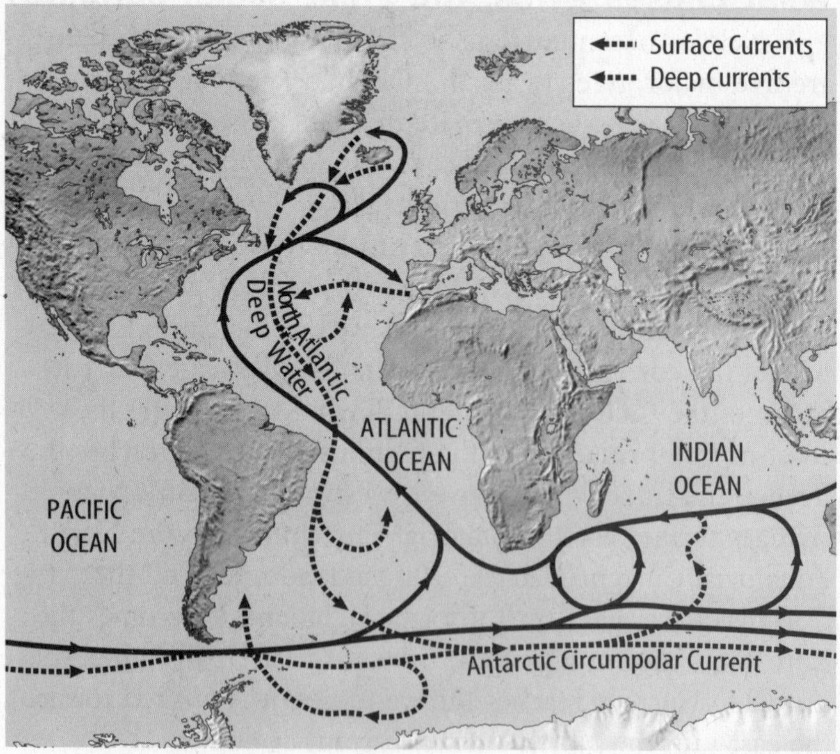

What are the effects of El Niño and La Niña on currents?

Recall that the trade winds drive the circulation of currents in the gyres. In the southern Pacific Ocean, they push tropical water westward from Central and South America toward Australia. Cool, deep water normally rises to the surface near South America. However, sometimes the trade winds weaken or even reverse direction.

When the trade winds stop driving the flow of water across the Pacific, the South Equatorial Current slows down. Warm water from the western side of the Pacific sloshes back across the ocean. This phenomenon is known as an El Niño (el NEEN yoh) event. Because ocean currents and winds are connected throughout the planet, El Niño conditions have effects all over Earth. These effects are felt strongly in California. During an extremely strong El Niño in 1997 and 1998, the rainfall in California was twice the normal amount. Landslides and avalanches occurred more frequently than usual.

When the trade winds begin to blow again, they usually do so with great strength. Warm tropical water is pulled across the Pacific toward Australia. The coast of South America becomes unusually cold and chilly. These conditions are called La Niña. El Niño and La Niña events occur about every three to eight years. Researchers are still trying to determine what drives these global-scale changes to the world's weather and ocean currents. ☑

Water Movement in the Ocean

Surface currents are driven by wind. The direction of surface currents is influenced by the Coriolis effect and land formations. Large gyres circulate in each major ocean basin. Deep ocean currents are driven by differences in water density. Ocean currents balance Earth's heat by transferring and distributing heat around the planet. Ocean currents also affect weather and climate. Warm-water currents, such as the Western boundary currents and the Gulf Stream, can influence regional climates by making them milder. El Niño and La Niña events, which act on winds and ocean currents, can influence weather all over the world.

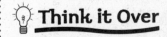

💡 Think it Over

6. Conclude What is the general direction of trade winds in the southern Pacific Ocean?

_____

_____

✔ Reading Check

7. Identify what happens during a La Niña event.

_____

_____

_____

chapter 10 Oceans

lesson 3 The Ocean Shore

Grade Six Science Content Standard. 2.c. Students know beaches are dynamic systems in which the sand is supplied by rivers and moved along the coast by the actions of waves.

MAIN Idea

The shore is shaped by the movement of water and sand.

What You'll Learn
- how waves shape the shore
- different types of sand

Mark the Text

Highlight Definitions
As you read, highlight the definitions of the underlined terms.

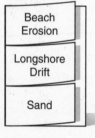

FOLDABLES

C Describe Make a three-tab Foldable. Label the front tabs as illustrated. Describe beach erosion, longshore drift, and sand in your own words under the tabs. Give specific example of how they shape the ocean shore.

> Beach Erosion
>
> Longshore Drift
>
> Sand

● Before You Read

Write about a time you put a lot of energy into splashing and making waves. Then read the lesson to learn how the movement of water and sand affects the ocean shore.

_____

_____

_____

● Read to Learn

Shoreline Processes

The <u>shore</u> is the area of land found between the lowest water level at low tide and highest area of land that is affected by storm waves. The <u>shoreline</u> is the place where the ocean meets the land. The location of the shoreline constantly changes as the tide moves in and out. Tides are the rising and falling of the surface level of the ocean. A beach is the area in which sediment is deposited along the shore. Beaches can be made of fine sand, tiny pebbles, or larger stones. The size and composition of the sediment that makes up a beach depends on where the sediment comes from.

What are the effects of wind and waves?

Wind and waves constantly beat the shoreline causing erosion. Wind picks up tiny pieces of sediment, called grit, and then smashes it against rocks. The grit acts like sandpaper, rubbing large rocks into smaller ones. Crashing waves force air and water into cracks in rocks, breaking them into pieces. Waves also hurl sand and gravel at the shoreline, wearing larger rocks down into smaller pieces. Finally, water itself can dissolve many minerals in rocks, causing them to break apart.

Erosion Shoreline erosion by wind and waves depends on two factors—the type of rock found in the area and the intensity of the wind and waves. Hard rocks, like granite and basalt, <u>erode</u> very slowly. Soft rocks, like limestone and sandstone, may wear away quickly.

Deposition Sediment that is eroded from one area of the shoreline eventually is deposited in another area. The deposition occurs where the waves are gentle. The sediment falls out of the water and settles on the seafloor. ☑

What happens when there is longshore drift?

Once sediments are eroded from rocks, they usually do not stay in one place very long. As waves approach the shoreline, they usually hit it at an angle. This process moves sand along the beach.

Part of the energy from the waves coming into the beach moves parallel to the shoreline. This energy drives a narrow current parallel to the shore called the **longshore current**. The longshore current can move up to 4 km/h. Longshore currents transport sand that is suspended in the surf along the shoreline, as pictured below. The combination of the movement of sand on the beach by breaking waves and the movement of sand in the longshore current is called **longshore drift**.

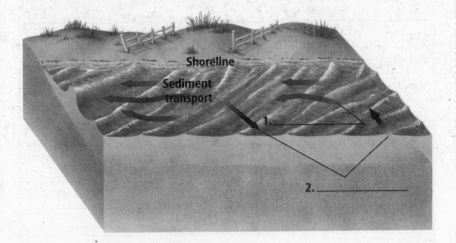

What are rip currents?

Usually the longshore current moves excess water along the beach. But when too much water piles up, the current cannot move it fast enough. The water breaks through the surf in a few places and rushes back out to the ocean. These swift currents that flow away from the beach are called **rip currents**.

✔ **Reading Check**

1. Describe Where is sediment deposited?

_____

_____

Picture This

2. Locate In the figure, write *Tidal current* by the arrows showing the movement of incoming and outgoing tides. Write *Longshore current* by the arrows showing this type of current.

How does human activity affect beaches?

To try to stabilize the beach, artificial structures often are put in place. Jetties, groins, and breakwaters all are structures that extend from the beach out into the water. Seawalls are built on land and usually are parallel to the shore. Sometimes building structures to protect beaches has unintended results. Breakwaters, jetties, and groins trap sand, which stops the normal flow of sand along the shoreline. Farther down the shoreline, the beaches may become smaller. The sand that would usually be deposited by longshore drift is trapped by the structures. Seawalls also can cause erosion. The wave energy that is deflected by the seawall can be redirected on either side and below it. This can erode sand from around the seawall, causing it to collapse. ☑

Sand and Weathered Material

Sand is a term that is used to describe rocks that are between 0.0625 mm and 2 mm. Within this range, sand is categorized as very coarse, coarse, medium, fine, and very fine, as shown in the table below. Rocks larger than sand are called gravel, cobbles, and boulders. Rocks that are smaller than sand are called clay and silt.

3. Explain How do artificial structures affect beaches?

_____

_____

_____

Picture This

4. Classify What kind of sediment is between 4 mm and 8 mm?

_____

Range of Sediment Sizes	
Name	**Size**
Boulder	> 256 mm
Cobble	64–256 mm
Very coarse gravel	32–64 mm
Coarse gravel	16–32 mm
Medium gravel	8–16 mm
Fine gravel	4–8 mm
Very fine gravel	2–4 m
Very coarse sand	1–2 mm
Coarse sand	½–1 mm
Medium sand	¼–½ mm
Fine sand	125–250 μm
Very fine sand	62.5–125 μm
Silt	3.90625–62.5 μm
Clay	< 3.90625 μm

Where does sand come from?

The effects of rain, freezing and thawing, and pounding by other rocks erode boulders away from their deposits. Weathering breaks large boulders into smaller rocks. Rain then washes small rocks into rivers. Rivers <u>transport</u> these rocks to the ocean. Along the way, the rocks are continually weathered and broken into smaller and smaller pieces. These small pieces of rock then are transported along the shoreline. Currents eventually deposit the broken up rocks as sand on sandy beaches.

What are the effects of sand deposits?

Even after sand reaches the ocean and then rests on a sandy beach, it does not stay there for long. Sand is continuously eroded, transported, and deposited along the shoreline. In the process, sand is sorted according to its size. The smaller grains of sand end up on low-energy beaches where the waves are gentle. The larger grains remain on high-energy beaches where waves are rough. Most sand usually ends up on the ocean floor.

Shaping the Shoreline

Beaches are areas of constant change. Wind and waves erode the rocks along the shoreline, creating features such as sea arches and sea stacks. Deposition of sediment occurs in areas of low energy, where sediment falls out of the water and settles on the seafloor. Human activities can affect the shape of beaches as well. Artificial structures such as breakwaters, jetties, and groins can interfere with the longshore current, resulting in abnormal erosion and deposition of sediment.

Academic Vocabulary

transport (TRAHNS port) (verb)
to carry or move from one place to another

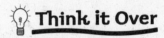 **Think it Over**

5. **Describe** What kind of waves likely strike a beach with fine sand?

_____

_____

Oceans

lesson ④ Living on the California Coast

Grade Six Science Content Standard. 1.e. Students know major geologic events, such as earthquakes, volcanic eruptions, and mountain building, result from plate motions. **Also covers:** 1.f.

MAIN ‹ Idea

Geology and ocean currents influence life in California.

What You'll Learn

- the geology of the California coastline
- how ocean currents affect California

Study Coach

Summarize As you read each paragraph, write a one or two sentence summary of its main ideas.

☑ Reading Check

1. **Explain** Why does California have so many rocky beaches?

_____

_____

_____

_____

● Before You Read

Have you ever been to the beach or seen a picture of one? Write a short description of the beach. Read the lesson to learn how the beaches of California were formed.

_____

_____

_____

● Read to Learn

Geology of the California Coast

The geology of California is based on the movement of tectonic plates. Most of California lies on the North American plate, while the Pacific Ocean rests on the Pacific plate. Until about 30 million years ago, these two plates smashed directly into each other. The force of this collision created the coastal mountains in Northern and Southern California. About 30 million years ago, the plates changed direction and started slipping past each other. This created a transform boundary. This slipping lifted up and crushed the sea floor into mountains in Central California. As a result of this tectonic activity, coastal mountain ranges stretch along the entire state of California.

What is the effect of high-energy waves on California's shoreline?

Very few offshore islands protect the coast of California. The waves that hit the shore carry a lot of energy with them. Because California rests on a transform boundary where plates are compressed, the shoreline is elevated. The high-energy of the waves erodes the cliffs along the shoreline, leaving large boulders and cobbles. ☑

What causes tsunamis?

Tsunamis are large sea waves. They are caused by anything that displaces a large amount of water, such as landslides, icebergs falling off glaciers, volcanic explosions, and undersea earthquakes. Undersea earthquakes create the largest tsunamis. ☑

When an earthquake occurs under the ocean, the movement of Earth displaces a large amount of water. This creates waves. The waves then move away from the location of the earthquake in all directions. When a tsunami reaches the shore, the bottom of the wave drags on the seafloor. This causes the enormous amount of water carried by the wave to pile up on itself. The excess water runs up on shore, similar to a fast and strong high tide.

Because there is tectonic activity throughout the Pacific Ocean, the California coast is at risk from tsunamis. Since 1812, 14 tsunamis have hit California, and 12 have caused damage. A sudden rise or drop in the sea level is a warning sign of an approaching tsunami. To stay safe, people should move to higher ground immediately. When tsunamis are generated far from California, the Tsunami Warning Center alerts local officials who will make decisions about evacuations.

Currents along the Coast

The **California Current** is a wide, slow-moving cold water current along the California coast. It flows in a southward direction, bringing cool water from northern latitudes. The **Davidson Current** is a narrow, warm water current that flows northward along the California coast. The two currents bump into each other near Point Conception, just north of Los Angeles. At this point, the California Current moves offshore. The Davidson Current is a seasonal current. It is stronger in the winter than the summer.

The chill in the ocean off California is caused by the California Current. The California Current pulls cold water from the northern latitudes along the California coast. The Davidson Current pushes it offshore.

What shields California from hurricanes?

The power of a hurricane comes from the warm water in tropical areas. When the storm system moves into cold water, it loses energy. Recall that the California Current brings cold water along the California coast from northern latitudes. The cold water along the coast acts as a hurricane shield.

✔ Reading Check

2. Describe How are the largest tsunamis created?

_____

_____

FOLDABLES

D Describe Make a three-tab Foldable. Label the front tabs as illustrated. Describe the California Current and the Davidson Current under the tabs. Explain what happens and what conditions exist when these two currents meet.

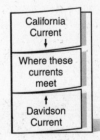

What is important in a sea life habitat?

The place in which an organism lives is called its <u>habitat</u>. Critical <u>elements</u> that affect a habitat include the types of food available, the shelter, the moisture, and the temperature ranges needed for survival.

In places where warm and cold water come together, the ocean is usually full of life. This is exactly what happens in the Channel Islands where the warm Davidson Current meets the cold California Current. The Channel Islands, along the coast of southern California, are home to a wide variety of different marine animals and plants. <u>**Marine**</u> refers to anything that is related to the ocean.

The rocky shore also provides a variety of habitats for the organisms that live there. The habitat on the rocky shore is often referred to as the intertidal zone. This is the area of shore that is between the lowest low-tide line and the highest high-tide line. The amount of water and exposure to air and the Sun varies in the intertidal zone. ☑

Coastal California

The tectonic activity of the North American Plate and the Pacific Plate has shaped the coast of California. Tectonic activity gave California its coastal mountains and rocky shorelines, as well as the threat of tsunamis. Water temperatures, sea life, and weather along the coast of California are influenced by the California Current and seasonal Davidson Current. The cold water carried by the California Current shields California from hurricanes. The diversity of marine life in the waters off of California, including the organisms living on beaches, rocky shores, and in coastal waters, can be affected by habitat change.

✔ Reading Check

3. Identify Which areas of the shore are part of the intertidal zone?

_____

_____

_____

_____

Weather and Climate

lesson ❶ Weather

Grade Six Science Content Standard. 4.e. Students know differences in pressure, heat, air movement, and humidity result in changes of weather. **Also covers:** 4.a.

● Before You Read

Think about the weather at your school today. Write a brief weather report on the line below. Be certain to include a description of the temperature, sun and cloud conditions, the wind, and if the air feels dry or damp. Read the lesson to learn more about describing weather.

_____

_____

● Read to Learn

Weather Factors

<u>Weather</u> is the atmospheric conditions and changes of a particular place at a particular time. If you have ever been caught in a rainstorm on what started out as a sunny day, you know that weather conditions can change quickly—over just a few short hours.

Temperature and rainfall are two ways you can describe weather. Barometric pressure, humidity, cloud coverage, visibility, and wind are other ways used to describe weather.

What affects air temperature?

Temperature is a measure of the average kinetic energy of the molecules in a material. When the air temperature is warmer the air molecules move faster. The temperature of the air is a measure of the average kinetic energy of air molecules.

Air temperatures change from morning to noon and into the evening. The winter season has cooler temperatures than summer. Temperature also depends on the distance from the equator where the sun shines directly. Temperature gets cooler as you go higher above the sea level, such as in the mountains.

MAIN ‹ Idea

Weather describes the atmospheric conditions of a place at a certain time.

What You'll Learn

■ the factors used to describe weather
■ the difference in the terms _humidity, relative humidity,_ and _dew point_
■ about the water cycle

Study Coach

Two-Column Notes As you read, organize your notes in two columns. In the left-hand column, write the main idea of each paragraph. Next to it, in the right-hand column, write details about it.

FOLDABLES™

Ⓐ **Define** Make a six-tab Foldable. Label the tabs as illustrated. Define each term under the tabs.

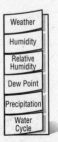

Weather
Humidity
Relative Humidity
Dew Point
Precipitation
Water Cycle

What is air pressure?

Air pressure is the force that a column of air exerts on the air below it. When you are close to the ground there is a lot of air above you pressing down. When you are high in the air there is not as much air pushing down. Therefore, air pressure is a larger number when it measured close to Earth's surface. At higher altitudes there is not so much air pressing down and the number for air pressure is lower. A barometer is the instrument used to measure air pressure. Air pressure is referred to as barometric pressure in a weather forecast. Knowing the barometric pressure of different areas helps meteorologists predict the weather. ☑

What part does wind play in weather?

Wind is air that is in motion relative to Earth's surface. Winds can blow from one direction and then change directions quickly. Wind can blow slowly and then quickly change to a fast wind at another time in the same day. Some winds, such as the westerlies and the trade winds, blow mostly in the same direction. Wind direction is given as the direction from which the wind is coming. You use the compass directions of north, south, east, and west to describe main directions for wind. For example, the westerlies blow from west to east. The polar easterlies blow from east to west.

What is humidity?

Water in the air is called water vapor. The amount of water vapor present in air is important for describing the weather. **Humidity** (hyew MIH duh tee) is the amount of water vapor per volume of air. You may have noticed that on days when the humidity is high there is more water vapor in the air. On a humid day, you might have noticed that your skin feels sticky and that sweat does not dry quickly. ☑

What does relative humidity mean?

When air is saturated it is full of water vapor. The amount of water vapor present in the air compared to the maximum amount of water vapor the air can hold at that temperature before becoming saturated is the **relative humidity**.

Relative humidity is given in percent. For example, a relative humidity of 50 percent means that the amount of water vapor in the air is one-half of the maximum the air can hold at that temperature. When weather forecasters give information about the humidity levels, they are usually referring to relative humidity.

☑ **Reading Check**

1. **Explain** Where is air pressure stronger, closer to Earth's surface or further away?

_____

_____

_____

☑ **Reading Check**

2. **Describe** What is humidity?

_____

_____

_____

What is the dew point?

The temperature at which air becomes fully saturated with water vapor and condensation forms is the **dew point**.

When the temperature drops, the air can hold less water vapor. The water vapor in air will condense to a liquid—dew, if the temperature is above freezing—or form ice crystals—frost, if the temperature is below 0°C. Warm air can hold more water vapor than cooler air. When the dew point is reached, the relative humidity is 100 percent. ☑

What causes clouds and fog to form?

When air reaches its dew point, water vapor condenses to form droplets. Clouds are water droplets or ice crystals that you can see in the air. Clouds can have different shapes and be present at different altitudes within the atmosphere. Since clouds move, they can move water and heat from one location to another. Recall that clouds are also important in reflecting some of the incoming solar radiation. When clouds form close to Earth's surface, it is called fog. Fog is a suspension of water droplets or ice crystals close to Earth's surface. Fog reduces visibility, which is the distance a person can see into the atmosphere. ☑

What is precipitation?

When water, in liquid or solid form, falls from the atmosphere it is called **precipitation** (prih sih puh TAY shun). Rain, snow, sleet, and hail are examples of precipitation. Rain is precipitation that reaches Earth's surface as droplets of water. Snow is precipitation that reaches Earth's surface as solid, frozen crystals of water. Sleet reaches Earth's surface as small ice particles that began as rain, but froze as they passed through a layer of below-freezing air. Hail reaches Earth's surface as large pellets of ice. Hail is formed from a small piece of ice that is repeatedly caught in an updraft within a cloud.

The Water Cycle

Water is essential for all living organisms. Approximately 96 percent of Earth's water is stored in the oceans. Fresh water is found in glaciers, polar ice, lakes, rivers, and under the ground and makes up only 4 percent of the water on Earth. The *hydrosphere* is the term used to describe all the water at Earth's surface. As shown in the figure at the top of the next page, water constantly moves between the hydrosphere and the atmosphere through the **water cycle**.

✔ Reading Check

3. **Define** What is the dew point?

_____

_____

_____

✔ Reading Check

4. **Explain** What is the difference between snow and sleet?

_____

_____

_____

Picture This

5. Identify Circle the process that occurs when water falls as rain, snow, or sleet.

_____

_____

_____

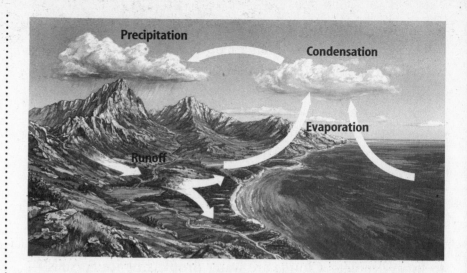

Precipitation

Condensation

Evaporation

Runoff

Academic Vocabulary

cycle (SI kul) (noun) a regular series of events or occurrences

Reading Check

6. Explain What is the importance of the sun's energy in creating weather changes?

_____

_____

_____

_____

_____

_____

How does the Sun affect the water cycle?

The Sun's energy drives the water <u>cycle</u>, as you can see in the figure above. Solar radiation that reaches Earth's surface causes water in the hydrosphere to change or evaporate from a liquid to a gas. Water that evaporates from lakes, streams, and oceans becomes water vapor in the Earth's atmosphere. As land and water is heated by the Sun, the air masses over them become warm and rise. As the air masses rise, the air expands and cools down. As the air cools down, the water vapor changes from a gas back into a liquid, a process called condensation. As the water vapor condenses, water droplets form. These water droplets then form clouds. Precipitation falls from the clouds to Earth's surface, returning water to the hydrosphere. ☑

Describing Weather and the Water Cycle

Temperature, precipitation, air pressure, and wind are some of the factors that are used to describe weather. The amount of water vapor in the air, or humidity, is also an important factor that determines the weather. You might feel sticky if the humidity is high on a hot day. If the air is saturated with water vapor, dew or frost may form as the air temperature drops.

Weather, along with the water cycle, is ultimately driven by the Sun's energy. As water moves between the hydrosphere and the atmosphere, water evaporates at Earth's surface, clouds form, precipitation falls, and water returns to Earth's surface. Precipitation may be in the form of rain, snow, sleet, or hail.

Weather and Climate

lesson ❷ Weather Patterns

Grade Six Science Content Standard. 4.e. Students know differences in pressure, heat, air movement, and humidity result in changes of weather.
Also covers: 2.d.

● Before You Read

Do you remember a time when the weather was very wet or very dry? Write a few lines about a weather event you have experienced or seen reported on television. Then read the lesson to learn more about weather patterns.

_____

_____

_____

● Read to Learn

The Changing Weather

Weather conditions can change rapidly. As high and low pressure systems are created, air masses move quickly from one area to another and create weather.

An **air mass** is a body of air that has the same weather features, such as temperature and relative humidity. Air masses get their characteristics from the surface over which they develop. For example, an air mass that forms over a warm dry area will become a warm dry air mass. As an air mass travels, it affects the local weather. For example, a cold air mass will bring cold temperatures.

What causes weather fronts?

The boundary between two air masses of different density, moisture, and temperature is called a front. As air masses move from one location to another, they eventually run into each other.

A **cold front** occurs when colder air moves toward warm air. The cold air pushes the warm air up into the atmosphere. The warm air cools as it rises and water vapor condenses. Clouds form and precipitation begins to fall. In many cases, cold fronts give rise to severe storms.

MAIN ⟨Idea

Many factors cause changes in weather.

What You'll Learn
- air masses and weather fronts
- high and low pressure systems
- severe weather events and their effects

◄ **Mark the Text**

Underline Unfamiliar Words As you read, underline any word that you do not know. Look for clues in the paragraph to the meaning of a word. Write the meaning of the underlined words in the margin.

FOLDABLES

❸ **Explain** Make a two-tab Foldable. Label the tabs as illustrated. Under the tabs, explain short- and long-term weather patterns in your own words and give examples of each.

Short-Term Weather Cycles

Long-Term Weather Cycles

How does a warm front develop?

A <u>warm front</u> forms when lighter, warmer air moves over heavier, colder air. Clouds form as the water in warm air condenses. A warm front usually results in steady rain for several days.

Have you ever heard a meteorologist using the terms *low-pressure system* and *high-pressure system*? Recall that when warm air rises, it creates a decrease in pressure close to Earth's surface. Therefore an area of low pressure is created. Areas of low pressure are associated with cloudy, stormy weather, and severe winds.

When cold air sinks, it moves closer to Earth's surface. The surface air pressure increases as air moves down, and an area of high pressure is created. This is called a high-pressure system. The sinking motion in high-pressure systems makes it hard for air to rise and for clouds to form. High-pressure systems are associated with fair weather.

What information does a weather map provide?

Information on weather factors, the presence of high-pressure and low-pressure systems, and weather fronts is usually represented in maps. These maps, called weather maps, provide useful information on the atmospheric conditions over areas of interest.

Weather maps contain a lot of information that is in the form of symbols. A key for each symbol is often placed next to the map. Weather maps usually provide information about temperature, precipitation, humidity, cloud cover, and pressure systems for an area.

Cycles that Affect Weather

Many cycles in nature affect weather. The cycles that regularly affect the weather include the day and night cycle, the seasons, and El Niño.

What are the effects of day and night cycles?

Air goes through a daily cycle of warming and cooling. As the Sun rises in the morning, sunlight warms the ground. The ground warms the air by conduction until a few hours past noon. As the Sun lowers in the afternoon, its energy is spread over a larger area. Sometime in late afternoon or early evening, the ground and air above begin to lose energy and start to cool. By late night or early morning, the coldest air is found next to the ground. ☑

Think it Over

1. **Predict** What kind of weather can you expect when a high-pressure system is present?

_____

_____

Reading Check

2. **Describe** When is the air nearest to the ground at its coldest?

_____

_____

_____

What are seasons?

As a result of Earth's tilt on its axis, the amount of solar radiation reaching different areas of Earth changes as Earth completes its yearly revolution around the Sun. **Seasons** are the regular changes in temperature and length of day that result from the tilt of Earth's axis, as shown below.

More solar radiation reaches the northern hemisphere during June, July, and August, resulting in summer. At the same time, less solar radiation reaches the southern hemisphere, where it is winter. During January, February, and March, less solar radiation reaches the northern hemisphere, resulting in winter. At the same time in the southern hemisphere, more solar radiation reaches Earth's surface and summer occurs.

During autumn and spring, neither pole is tilted toward the Sun.

23.58

Autumn

During the northern hemisphere summer, the north pole is tilted toward the Sun.

Summer

147,000,000 km   Sun   152,000,000 km

Winter

During the northern hemisphere winter, the south pole is tilted toward the Sun.

Spring

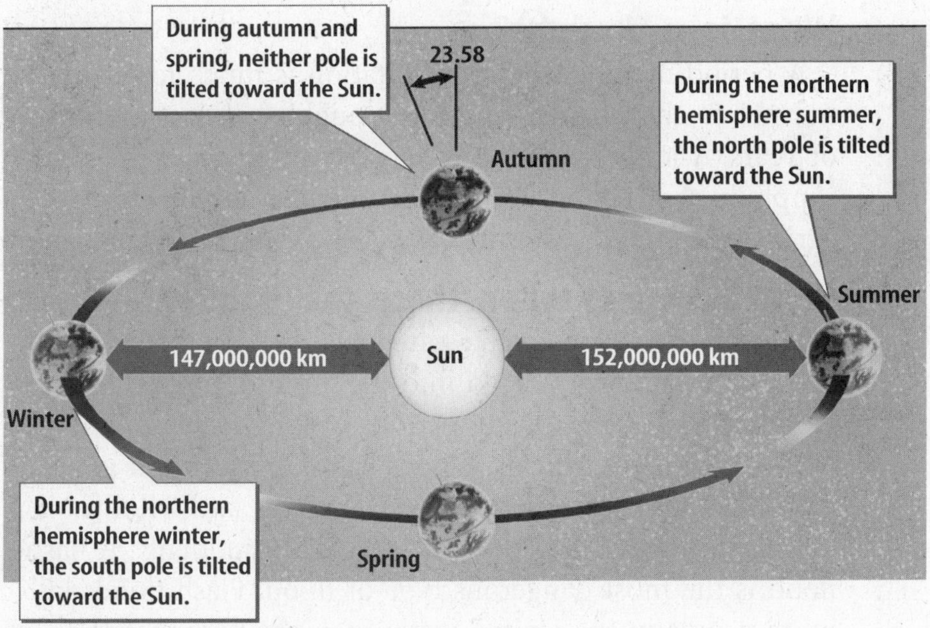

What is El Niño?

The alternating cycles of El Niño and La Niña can affect weather worldwide. During El Niño, the surface water is warmer in the waters of the eastern Pacific Ocean near the equator. The warmer water causes more water vapor in the air above the water. ☑

The main effects of El Niño include heavy rainfall and flooding in California, the southeastern United States, and the South American countries of Peru and Ecuador. Winter temperatures in the north central United States are usually warmer than normal. El Niño also tends to result in fewer hurricanes in the Caribbean and southeastern United States. However, El Niño often produces severe droughts that can lead to forest fires in Australia, Indonesia, and southeast Africa.

✔ Reading Check

4. Explain the effect warm water has on the air above it during El Niño.

_____

_____

_____

What are the effects of La Niña?

A La Niña event occurs when the surface water in the eastern Pacific Ocean near the equator is colder than normal. The effects of La Niña are winter temperatures in the northwestern United States that are colder than normal. Winters in the southeastern United States are often warmer than normal during La Niña.

Severe Weather

Sometimes weather can be severe and cause dangerous conditions. For example, heavy rainfall can lead to floods. Other times, rainfall can be absent for long periods of time. These events can kill humans, animals, and plants.

What is a drought?

A period of time when precipitation is much lower than normal or absent is a **drought**. Droughts can last months or years. A long drought in a region can decrease the water supply in that region. A decrease in water supply affects agriculture and, in extreme cases, can result in famine.

What happens when there are floods?

Floods occur when water enters an area faster than it can be taken away by rivers, go into the ground, or into lakes. Flooding can occur during periods of long heavy rain or when snow melts quickly. Floods can damage, wash away, or even bury the living areas of both humans and wildlife.

A **flash flood** is a flood that takes place suddenly. A flash flood is the most dangerous type of flood. Flash floods kill more people in the United States than any other weather-related cause of death.

Human activities can lead to more damaging flash floods. This is because the construction of buildings, parking lots, and other structures decreases the amount of vegetation and soil that can potentially absorb runoffs of water. ☑

Changes in Weather

Cold fronts and warm fronts lead to changes in temperature and precipitation. Low-pressure systems can bring rain and high-pressure systems can bring fair weather. Weather is affected by daily cycles of warming and cooling. Weather is also affected by longer cycles, including seasonal changes and El Niño and La Niña events. Weather cycles can cause severe weather such as droughts or floods.

💡 Think it Over

5. Evaluate Which weather pattern seems to cause the more weather disasters in the United States? (Circle your answer.)
a. El Niño
b. La Niña

☑ Reading Check

6. Conclude Why are flash floods the most dangerous type of flood?

_____

_____

_____

Weather and Climate

lesson ③ Climate

Grade Six Science Content Standard. 4.d. Students know convection currents distribute heat in the atmosphere and oceans.

● Before You Read

Think about the weather during a year in your town or city. Write a sentence that tells if your home area is mostly cool or warm. Then read the lesson to learn about climates.

_____

_____

● Read to Learn

A World of Many Climates

You live in a world of different climates—from hot and dry desert areas, to warm and wet rainforest regions, to frigid cold tundra at the poles. The climate of different regions affects all of the organisms living there

What is climate?

<u>Climate</u> is the long-term average of the weather patterns of an area. Climate includes temperature, winds, and precipitation of a long period of time.

What are climate regions?

The different climates of Earth are classified into climate regions. One popular classification system uses temperature, precipitation, and vegetation as characteristics to describe the climate of a region. ☑

The main climate regions of Earth include:
- cold polar tundra,
- dry desert,
- mediterranean,
- humid subtropical,
- highland,
- humid continental, and
- marine.

MAIN Idea

The climate of a region is often defined by annual temperatures and precipitation amounts.

What You'll Learn
- the characteristics of a mediterranean climate and a highland climate
- ways that human activities can affect the climate

Study Coach

Outline Make an outline to organize the information in this lesson. The numbered headings should be the main points in the outline. Under each heading, list the details or examples that help explain the main idea.

✔ Reading Check

1. Identify What three characteristics are used to classify climate regions?

_____

_____

_____

Earth's Climate Regions

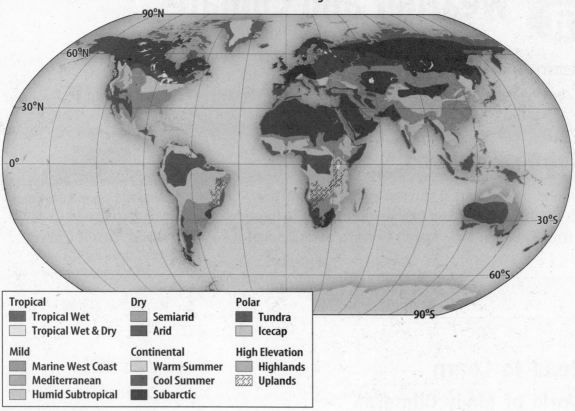

Tropical
- Tropical Wet
- Tropical Wet & Dry

Dry
- Semiarid
- Arid

Polar
- Tundra
- Icecap

Mild
- Marine West Coast
- Mediterranean
- Humid Subtropical

Continental
- Warm Summer
- Cool Summer
- Subarctic

High Elevation
- Highlands
- Uplands

Picture This

2. Classify Is your region composed of one climate or a mixture of climates?

_____

FOLDABLES™

C Compare Make a Venn-diagram Foldable. Record what you learn about mediterranean and highland climates as they relate to California under the appropriate tabs.

What are California's climates?

Most of California has mediterranean and highland climates. A **mediterranean** (me dih tur RAY nee uhn) **climate** is characterized by mild, wet winters and hot, dry summers. Mediterranean climates usually occur on the western side of a continent. Because mediterranean climates are often extremely dry, summer fires often occur in those regions. A **highland climate** is characterized by cool-to-cold temperatures and occurs in the mountains and on high plateaus.

Climate Controls

Uneven heating of Earth's surface by the Sun results in air currents and ocean currents that influence the different climate regions. Other factors, such as the latitude and altitude of a location, the distance from a large body of water, and the presence or absence of mountain barriers, also affect the climate experienced in different areas. These factors are called climate controls. Some of these factors, such as latitude and mountain barriers are unchanging. Other factors, such as winds and ocean currents, can vary according to seasons.

What do you need to know about latitude?

Recall that areas close to the equator receive more solar radiation than areas located further north or south. Since more solar radiation is received in areas near the equator, these areas have warmer climates than those regions at higher latitudes. ☑

How does the distribution of land and water affect climate?

The distribution of land and water has an important influence on climate. Recall that water can absorb or lose large amounts of heat without changing temperature. Land does not have this characteristic. Land surfaces heat and cool rapidly. Ocean surfaces heat and cool slowly. As a result, the climate of locations near an ocean is affected as illustrated in the map below. For example, San Francisco and Wichita are cities located at the same latitude. However, they do not have the same climates. San Francisco is warmer in winter and cooler in summer because it is near the ocean.

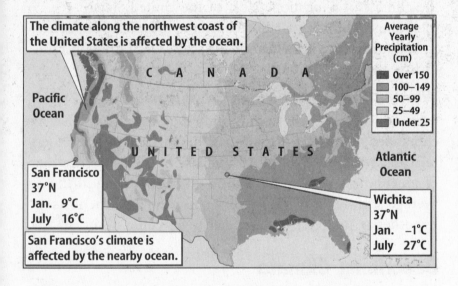

The climate along the northwest coast of the United States is affected by the ocean.

Pacific Ocean

San Francisco
37°N
Jan. 9°C
July 16°C

San Francisco's climate is affected by the nearby ocean.

Average Yearly Precipitation (cm)
- Over 150
- 100–149
- 50–99
- 25–49
- Under 25

Atlantic Ocean

Wichita
37°N
Jan. –1°C
July 27°C

How do ocean currents affect climate?

Ocean currents help to redistribute the Sun's energy on Earth. Ocean surface water near the equator receives more solar radiation than surface water at high latitudes. Large ocean currents that move water away from the equator, such as the Gulf Stream, carry heat to higher latitudes. Ocean currents moving toward lower latitudes, such as the California Current, replace the warm water from the lower latitudes with cold water from higher latitudes.

3. Explain Why do areas near the equator have warmer climates?

_____

_____

_____

Picture This

4. Explain Study the map. Why is San Francisco warmer in winter and cooler in summer than Wichita?

_____

_____

_____

How do the Gulf Stream and California Current affect climate?

This redistribution of heat by ocean currents can be seen when comparing the average winter temperatures of Great Britain to those of Labrador, Canada. These two locations are found at the same latitude. However, due to the influence of the warm Gulf Stream, the average winter temperature in Great Britain is 15°C–20°C warmer than that of Labrador. The cold Labrador Current runs along the coast of Labrador, keeping winter temperatures low.

The California Current is a cold-water current that begins in the North Pacific Ocean and flows past the coast of California. The water temperature of the California current remains <u>stable</u> year-round. This helps keep temperatures along the coast from rising too high in the summer and from dipping too low in the winter.

What are prevailing winds?

A prevailing wind is a wind that blows most often across a particular area. Prevailing winds significantly influence the climate of a location. Most of the United States, including California, is affected by the prevailing westerlies. In California, the influence of the westerlies results in rain during the winter months.

What effect do humans have on climate?

Some scientists believe that human activities can affect the climate of our planet. Recall that the burning of fossil fuels releases greenhouse gases. An increase in the concentration of greenhouse gases could lead to global warming. If the average surface temperature of Earth increases, scientists hypothesize that global climate changes could occur. ☑

Describing Climate

Climate is usually described in terms of the average weather conditions of an area over a long period of time. Conditions such as temperature, precipitation, and winds are taken into account when describing a climate. Most of California has either a mediterranean climate or a highland climate. Factors that influence climate include latitude, altitude, distance from a large body of water, and presence or absence of mountain barriers. The climate of the coastal areas of California is influenced by the California Current. Altitude or mountain barriers may influence climate in inland areas of California.

Academic Vocabulary
stable (STAY bul) (adj)
unchanging; steady

☑ **Reading Check**

5. Name one human activity that might affect climate.

_____

_____

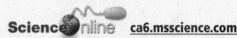

Weather and Climate

lesson ❹ California Climate and Local Weather Patterns

 Grade Six Science Content Standard. 4.e. Students know differences in pressure, heat, air movement, and humidity result in changes of weather.
Also covers: 4.d.

● Before You Read

What kind of outdoor sports can you enjoy where you live because of the weather? Write a sentence about the sports that are popular in your area. Read the lesson to learn about California's climate.

_____

_____

● Read to Learn

Mediterranean and Highland Climates

The mediterranean climate of California is influenced by westerlies in winter. The Pacific Ocean and the mountains also influence the climate of California.

Locations in California that are at high altitudes have highland climates. The temperatures in these regions are lower than other regions with lower altitudes. Precipitation is generally greater in highland areas than in locations at lower altitudes.

What are the seasonal changes in California?

Although winters in California are mild, there is a contrast between summer and winter in California. This contrast is best described in terms of precipitation. Instead of referring to seasons in California as hot and cold, the terms _rainy_ and _dry_ often are used. Dry summers and wet winters are characteristic of the mediterranean climate of California. California's dry season also is called the fire season. Less rain occurs in summer due to the presence of an offshore high-pressure system referred to as the Pacific High. Rainfall and other precipitation in California is higher from November to March. ☑

MAIN ❰ Idea

California's climate is primarily mediterranean and highland.

What You'll Learn

■ how rain shadows are developed
■ to compare the winds off the sea, land, mountain, and valley

---

▸ Study Coach

Identify Main Ideas As you read, write one sentence to summarize the main idea in each paragraph. Use vocabulary words in your sentence.

✔ Reading Check

1. **Describe** What are the characteristics of the mediterranean climate of California?

_____

_____

_____

Where are California's fog belts?

Much of the Pacific Coast of California is known for the presence of fog. During summer, warm air masses <u>accumulate</u> far offshore. These warm air masses contain a lot of moisture since they are in contact with the ocean and their temperatures are high. The westerlies move the air masses and the moisture from the west to the east. As this happens, the air masses moving to the coastal areas are cooled when they cross over the cold California Current. When the air masses cool down, water condenses, giving rise to fog.

This fog is essential to the survival of the redwoods on the northern coast of California. The redwoods are able to survive the dry summer by taking in up the water they need from the fog that rolls in every day during the summer months. The fog collects on the needles and branches, drips to the ground, and is absorbed by the tree's shallow root system.

What are the California's rain shadows?

There are areas in California where rainfall is low. One reason for this is the presence of rain shadows. An area of low rainfall on the slope of a mountain that is away from the wind is called a **rain shadow**. Warm westerly breezes bring warm, moist air inland across California. When the warm damp air runs into mountains, the warm air rises. When the air rises, it cools down and begins to rain or snow. The air that reaches the other side, or downwind slope, of the mountain is dry. As a result, the area on the downwind slope of the mountain is a rain shadow that experiences low rainfall.

Local Winds

Recall that winds blow from regions of high pressure to regions of low pressure. Local changes in pressure can give rise to local winds such as sea breezes, land breezes, valley breezes, and mountain breezes.

What are sea breezes and land breezes?

If wind is blowing from the sea to the land it is called a **sea breeze**. A wind that blows from the land to the sea is called a **land breeze**. ☑

Sea breezes and land breezes are caused by atmospheric convection currents. Solar radiation heats the land and ocean. However, the temperature of the land changes more rapidly than the temperature of the ocean. When the air above land is heated, it rises and creates an area of low pressure.

Academic Vocabulary
accumulate (ah KYEW muh layt)
(verb) to gather or collect over time

FOLDABLES

D Define Make a six-tab Foldable. Label the tabs as illustrated. Define each term under the tabs.

Rain Shadow
Sea Breeze
Land Breeze
Mountain Breeze
Valley Breeze
Santa Ana Wind

☑ Reading Check

2. **Distinguish** Which breeze blows from the sea to the land? (Circle your answer.)
   a. sea breeze
   b. land breeze

How does air pressure affect sea breezes and land breezes?

In the case of the sea breeze, the area over the ocean has a higher pressure than the area over land where air is rising. Sea breezes usually blow from the ocean to the land during the daytime.

Land breezes are more common during the nighttime. At night, land cools faster than the ocean. The air sinks over the land and makes high pressure compared to the area over the ocean. As a result, a land breeze blows from the land to the ocean, as illustrated below.

Picture This

3. Interpret What is happening to the warm air in both figures?

_____

_____

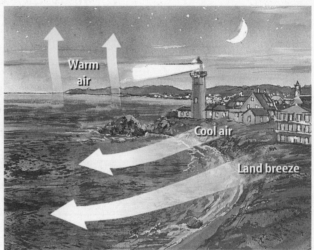

What is a valley breeze?

A valley breeze is named for the direction from which it blows. **Valley breezes** blow from the valley along the mountain slope. Air heated during the day in the valley causes warm air to rise along the slopes of the mountains. Usually valley breezes occur during daytime hours when the land is hot.

What is a mountain breeze?

Like a valley breeze, a mountain breeze is named for the direction from which it blows. **Mountain breezes** flow from the mountaintop downward. Mountain breezes usually occur at night in wide mountain valleys that were heated by the Sun during daylight hours. After sunset, air along the exposed mountain slopes cools more rapidly than the air in the valleys. The cooler mountain air sinks into the valley. ☑

☑ **Reading Check**

4. Determine When do mountain breezes flow?

_____

_____

_____

What are the Santa Ana winds?

In southern California, hot, dry winds that blow from the east or northeast and continue toward the coast are often called **Santa Ana winds**. The Santa Ana winds blow from the mountain passes and canyons of southern California across the Los Angeles basin. One of the canyons through which the wind blows is the Santa Ana Canyon from which the wind gets its name.

Santa Ana winds begin as cool, dense, and dry air. The dense air is compressed and warmed as it is forced down through narrow canyons. By the time the Santa Ana winds reach coastal areas they have high speeds and high temperatures. ☑

Fires Winds can be hazardous during fire season. Winds that are hot, dry, and moving fast, can dry out vegetation and increase the danger of fire. The names and descriptions of the winds that contribute to the spread of fires in California are shown in the table below.

☑ **Reading Check**

5. **Describe** What are the Santa Ana winds like when they reach the coastal areas?

_____

_____

Significant Local California Winds During Fire Season	
Wind Name	**Characteristic and Facts**
Santa Ana wind	• blows from Great Basin into Los Angeles Basin and San Fernando Valley • named for blowing through Santa Ana Canyon
Mono wind	• blows from the Sierra Nevadas into the Great Central Valley • named for blowing from the direction of Mono Lake
California norther	• blows from northern California into the northern half of Great Central Valley • brought temperature in Red Bluff to 119°F, August 8, 1978
Diablo wind	• blows from the east into San Francisco and Oakland • The costliest fire on record was in the Oakland hills, October 1991.
Sundowner wind	• blows from Santa Ynez Mountains into Santa Barbara • named for blowing during late afternoon or early evening at sundown

Picture This

6. **Identify** Which wind was named for the time of day it blows?

_____

California's Climate

California has two main climate regions—mediterranean and highland. Differences in precipitation and temperature define these two regions. Coastal areas of California experience fog, sea breezes, and land breezes. Valley breezes and mountain breezes blow up and down the sides of mountains. The downslope of mountain ranges are in rain shadows and experience low amounts of rainfall. Different winds that bring hot, dry air to an area can influence temperatures and increase the risk of fires.

Ecological Roles

lesson ❶ Abiotic and Biotic Factors

Grade Six Science Content Standard. 5.e. Students know the number and types of organisms an ecosystem can support depends on the resources available and on abiotic factors, such as quantities of light and water, a range of temperatures, and soil composition.

● Before You Read

Rocks and soil are both nonliving parts of a garden. On the lines below, describe how they affect the plants that grow. Then read on to find out more about Earth's ecosystems.

_____

_____

_____

_____

MAIN ❮Idea

Living things and nonliving factors interact in Earth's ecosystems.

What You'll Learn

- how organisms in an ecosystem depend on abiotic factors
- how systems depend on biotic and abiotic factors
- how changing abiotic factors can cause organisms to adapt

● Read to Learn

What is an ecosystem?

An **ecosystem** consists of the organisms in an area and the physical place they live. Organisms are living things. Earth has many types of ecosystems. Oceans, deserts, parks, and gardens are examples of ecosystems. The organisms that share an ecosystem interact with each other. They also interact with nonliving factors in their environment.

Abiotic Factors

Abiotic factors are the nonliving parts of an ecosystem. Sunlight, soil, water, and air are important abiotic factors. They help control the growth of plants.

Why is the Sun an important abiotic factor?

Most life on Earth depends on energy from the Sun. For example, green plants use the Sun's energy to make food. Other organisms eat those plants. Organisms also depend on heat from the Sun to keep warm. The Sun's energy controls many of the abiotic factors in our environment.

Study Coach

Define Terms As you read, write each key term on an index card. Write the definition on the back of the card. Use the cards to review important words after you finish reading the section.

FOLDABLES™

Ⓐ Compare and Contrast Make a two-tab Foldable and label the tabs as illustrated. Record information about biotic and abiotic factors under the tabs and use what you learn to compare and contrast each.

Biotic Factors | Abiotic Factors

✔ **Reading Check**

1. **Explain** Why do different areas of Earth receive more or less radiation from the Sun?

_____

_____

_____

Think it Over

2. **Predict** What would happen if all precipitation occurred only over the oceans?

_____

_____

How does the Sun affect climate?

Climate is another abiotic <u>factor</u>. It is the pattern of weather in an area over many years. Temperature, wind, precipitation, and sunlight are all part of an area's climate.

Earth's surface is curved. As a result, different areas of Earth receive different amounts of radiation from the Sun. Sunlight hits the north and south poles at an angle. This spreads the heat, causing the poles' climate to be very cold. Few species are able to live there. However the equator is hit directly by the Sun's rays. This creates some of the warmest places on Earth. Life is abundant at the equator's regions. Temperature differences between the poles and the equator also affect wind and wave patterns around the world. ✔

What is the water cycle?

Water is an abiotic factor that is important to almost all life on Earth. Earth itself is between 70 and 75 percent water. Most of Earth's water is in oceans. Much of the rain or snow that falls on land drains into a river or stream. Eventually it flows into the sea. The process of getting water from the sea to the sky and back to the land is called the water cycle. It requires energy from the Sun. The water cycle is described below.

What are the steps in the water cycle?

The following steps repeat in an endless cycle. The water cycle makes water available to support life on land.

1. Water from oceans, rivers, and other bodies of water is evaporated by sunlight.
2. Water molecules condense in the sky and form clouds.
3. Water from the clouds falls back to Earth as precipitation, such as rain, snow, sleet, and hail.
4. Some of the water falls back into oceans and other bodies of water. The rest falls onto land.
5. Some of the water that falls onto land is used by plants and animals. Some soaks into the ground and flows into nearby bodies of water.

What type of soil do most plants need?

Soil is an abiotic factor that is essential for many plants. Soils are not all the same. For one thing, soil can contain minerals such as limestone and quartz. These <u>affect</u> the soil's acidity. Plants grow poorly in soil that has too much or too little acid. Most farmers monitor the acidity of their soil.

Soil also contains nutrients plants need. Three of these nutrients are nitrogen, phosphorus, and potassium. Plants grow best in soil that has the right amount of each. Nutrients are added to soil when dead plants and animals decay in it. The decaying matter is called **humus** (HYEW mus). Humus lies in a thin layer on the soil's surface. As it decays, it adds nutrients to the soil. This helps plants grow. Humus also helps plant growth by soaking up water, which keeps the soil moist. In nature, regions that have the most humus in the soil also have the most plant growth. ☑

How does air support life on Earth?

Humans and many other organisms need to live in places where the air contains enough oxygen. This is because our cells need oxygen to release energy. Cells in organisms begin to die after only five minutes without oxygen. Some organisms need less oxygen than others. The air in a particular ecosystem helps determine the organisms that will live there.

Living things affect air too. Some organisms, like humans, take oxygen from the air and release carbon dioxide while breathing. Other organisms, such as plants, take in carbon dioxide and release oxygen.

Biotic Factors

Every ecosystem is made of living and nonliving parts. <u>Biotic factors</u> are the living parts of an ecosystem. Plants, animals, and all other living things are biotic factors. Biotic factors interact with abiotic factors to determine the numbers and types of organisms in an ecosystem.

How do living things interact?

In an ecosystem, all organisms depend on each other. Think about a coral reef ecosystem. Corals are tiny animals. Microscopic algae live inside the coral. The algae provide food and oxygen for the coral. The coral provides a place for the algae to grow. Together, they make large reefs that can be home to snails, shrimp, lobsters, and other organisms. Plants also grow in the reefs. Those plants provide hiding spaces for small fish.

Academic Vocabulary
affect (a FEKT) (verb)
to influence

✔ **Reading Check**

3. **Describe** Name two ways that humus is helpful to plants.

_____

_____

💡 **Think it Over**

4. **Apply** What are three biotic factors that you might find in a forest?

_____

_____

💡 **Think it Over**

6. Determine What is a limiting factor in your environment?

_____

_____

_____

Different Roles In a coral reef, like all ecosystems, living things compete for food and space. Some help each other by providing food, a place to live, or a place to hide. A **species** is a group of organisms that share similar characteristics and can reproduce among themselves and produce fertile offspring. Each species plays a different role. Together they make up the biotic factors of the ecosystem. ✔

How does a population differ from a community?

Imagine a pond with frogs living in it. If it contained thirty frogs then it would have a frog population of thirty. A **population** is the number of individuals of one species that live in the same area. A **community** is all the species that occupy an area. In the case of a pond, the community includes all species of plants and animals that live in and around the pond.

Limiting Factors

In nature, populations expand until limited by biotic or abiotic factors. A **limiting factor** is something in the environment that limits how large a population can grow. Food, water, space, and shelter are common limiting factors.

Limiting factors can change over time. If an ecosystem that gets lots of rain suddenly has a drought, water becomes the limiting factor. Lack of water could cause some individuals to die. Different factors limit different species. Sun-loving plants do not grow well under large trees. Access to sunlight might be a limiting factor for those plants.

Changes in Population

The life of sea otters can show how abiotic and biotic factors affect ecosystems. Sea otters live in kelp beds. Kelp are giant algae that attach to the ocean floor. Sea urchins chew off the kelp where they are attached. Sea otters eat sea urchins. When the sea otter population declines, the sea urchin population increases and destroys the kelp beds. The populations of other organisms that depend on kelp for shelter and food, such as clams, snails, and octopuses, also decline.

The sea otter population is not increasing as fast as scientists hoped. One reason may be that killer whales are eating more sea otters. The whales' usual food sources, seals and sea lions, have declined in recent years. Some scientists hypothesize that warmer ocean temperatures, an abiotic factor, and over-fishing may be responsible for the decline in seals and sea lions.

Ecological Roles

lesson 2 Organisms and Ecosystems

Grade Six Science Content Standard. 5.c. Students know populations of organisms can be categorized by the functions they serve in an ecosystem. **Also covers:** 5.d., 5.e.

● Before You Read

Think about the differences between a desert and a forest. Write some of those differences here. Then read the lesson to find out more about Earth's biomes.

_____

_____

_____

● Read to Learn

Biomes

A **biome** is a large geographic area that shares a similar ecosystem and climate. Each biome has its own climate and type of soil. Different biomes also contain different organisms. The map on the next page shows the major biomes found on Earth.

What are characteristics of the tundra?

The tundra biome is located near the north and south poles. During the winter it receives little water or sunlight. The ground can stay frozen all year.

The tundra supports more organisms during its short summer than in the cold, dark winters. Many animals that live in the tundra in the summer migrate to warmer climates in the winter. Animals that stay either hibernate or are adapted to the cold and dark.

What is the taiga?

The taiga (TI guh) is a cold, forest region with many evergreen trees. It is the largest biome on Earth. The taiga is warmer and wetter than the tundra, though winters are long and cold. The taiga biome stretches across North America, northern Europe, and Asia.

MAIN Idea

Each of Earth's biomes has its own climate and supports a variety of organisms.

What You'll Learn
- about different biomes
- how different organisms play similar roles in different ecosystems

---

Mark the Text

Identify Main Ideas Read each of the question heads. As you read the paragraphs related to each head, underline the answer to the question.

FOLDABLES

B **Take Notes** Make a Foldables chart and label the three columns as illustrated. Use the chart to take notes on what you learn about biomes, climate, and community.

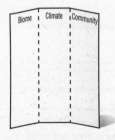

What are the two types of rain forests?

There are two types of rain forests. Both receive abundant rainfall and are home to many lush, green plants. As shown on the map below, tropical rain forests are found near the equator. Tropical rainforests have more plant and animal species than any other biome. Temperate rain forests are found in coastal regions that receive much rainfall each year.

What biomes are most common in California?

Several biomes are found in California. The two major biome types in California are the desert and the grasslands. There is also a biome in California called the temperature deciduous forest. ☑

✔ Reading Check

1. **Identify** What are the two major biomes in California? (Circle your answer.)
   a. desert and rain forest
   b. desert and grasslands

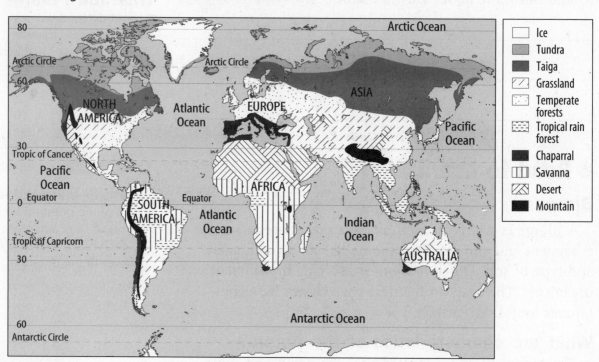

Picture This
2. **Locate** Circle the names of continents on which deserts are found.

Temperate Deciduous Forests Plants in a temperate deciduous forest include trees that lose their leaves in the winter. Some evergreens also grow in these forests. Animals found in California's deciduous forests include insects, rodents, foxes, deer, raccoons, bobcats, and wolves. California's temperate deciduous forests are found in the northern part of the state, which has abundant rainfall in the summer and cold winters.

Deserts The desert has less than two centimeters of rainfall each year. Life in the desert has adapted to conserve water. Cacti have tough, waxy coatings with thick leaves that store water. Animals such as rodents and snakes live in California's deserts. Common desert plants are shown on the next page.

Grasslands Most grasslands have a dry season each year with little or no rain. This lack of moisture prevents forests from developing. Grasslands have different names. In North America, they are called prairies. In Africa, grasslands are known as savannahs. Southern California's coastal <u>regions</u> are chaparral biomes. The chaparral biome has hot, dry summers. The winters are mild and rainy. Fires and droughts are common. Chaparral plants and animals are adapted to these conditions.

Some chaparral organisms depend on fire. For example, the seeds of some pine trees are sealed in resin. When there is a hot fire, the resin is melted and the seeds can grow. The picture below shows the chaparral shortly after a fire.

Picture This
3. Predict Where will desert animals find shelter?

_____

_____

_____

_____

Academic Vocabulary
region (REE jun) (noun) a broad geographic area distinguished by similar features

Picture This
4. Identify Study the photo of the chaparral biome. What evidence is available to prove that fire once destroyed the area?

_____

_____

_____

Habitat and Niches

Each species fills a different role in the community and needs different things to survive. The place an organism lives is called its **habitat**.

What is a niche?

The role that an organism plays in its environment is called its **niche** (NICH). A niche describes how an organism interacts with all the biotic and abiotic factors in its environment.

Every ecosystem has similar niches. However, niches can be filled by different species in each ecosystem. For example, all ecosystems have niches for predators. In South American rain forests, those niches are filled by animals such as harpy eagles, monkeys, and jaguars. In your own back yard, similar niches are filled by robins and housecats, as illustrated below.

Academic Vocabulary
environment (ihn VI run munt)
(noun) circumstances, objects, or conditions by which one is surrounded

Picture This

5. Predict Circle the organism in this image that might fill the niche of "predator".

Human Impacts on Niches

What would happen if an organism was no longer able to fill its niche in an ecosystem? The entire ecosystem would likely be affected by this change. A population decline of one type of organism affects the rest of the ecosystem.

Human activities affect population size and have an impact on entire biomes. For example, if people did not allow fires to burn in chaparrals, there would be no young pine trees. As older pine trees died, they would not be replaced. The animals that depend on these trees for food and shelter would either have to move or die.

Think it Over

6. List two ways that human activities might affect biomes.

_____

_____

_____

_____

Science online   ca6.msscience.com

Energy and Matter in Ecosystems

lesson ❶ Producers and Consumers

Grade Six Science Content Standard. 5.c. Students know populations of organisms can be categorized by the functions they serve in an ecosystem.
Also covers: 5.a, 5.e.

● Before You Read

What kinds of foods do you eat? Do they come from plants, animals, or both? On the lines below, describe your breakfast or lunch. Then read the lesson to learn more about how energy from the Sun is converted into food.

_____

_____

MAIN ⟨Idea

Producers make their own food. All other organisms depend on producers as their source of energy.

What You'll Learn
■ how to classify consumers as herbivores, carnivores, and omnivores

● Read to Learn

Ecosystems

An ecosystem is a community of plants and animals and the environment in which they live. Each ecosystem is made up of biotic and abiotic factors. Biotic factors are the living things. Abiotic factors are things such as water, sunlight, and soil type. Abiotic factors determine the types of organisms that will be able to live in an ecosystem.

What is ecology?

<u>Ecology</u> (ih KAH lu jee) is the study of the ways living things interact with each other and with their environment. One part of ecology is studying the ways in which energy and matter move through an ecosystem. Ecology also involves studying all the species and communities that are in an ecosystem.

Producers

<u>Producers</u> are organisms that use energy from the Sun or other chemical reactions to make their own food. Green plants, algae, and some microorganisms are producers. Most use the energy from sunlight. However there are some bacteria that use energy from chemical reactions instead.

Study Coach

Preview Headings Read each of the question headings. Answer the question using information you already know. Then read the section and answer the question again.

FOLDABLES

Ⓐ Record Information
Make four note cards. Use two note cards to record what you learn about the importance of producers ad consumers to an ecosystem. Make extra not cards to record and define new terms or to explain new concepts.

What is photosynthesis?

<u>Photosynthesis</u> (foh toh SIHN thuh sus) is a process in which producers use energy from sunlight to make their own food. This is the most common way in which energy and carbon enter the web of life. The energy from sunlight is used to turn carbon dioxide and water into simple sugars. Those sugars are the producers' food. The figure below shows the chemical equation for photosynthesis.

Picture This

1. **Identify** Circle the three things needed for photosynthesis to take place.

Photosynthesis

$$6CO_2 + 6H_2O + \text{light energy} \longrightarrow C_6H_{12}O_6 + 6O_2$$

carbon dioxide   water   chlorophyll   sugar   oxygen

How do plants grow?

Green plants make their own food through photosynthesis. They also need nutrients. Plants take in nutrients from the soil. The nutrients are combined with carbon from simple sugars. These are used to make starches, proteins, oils, and other compounds. These compounds are used to build roots, stems, leaves, and seeds.

What other organisms are producers?

Not all producers are green plants. Algae are producers too. Algae are protists that make their own food. This is done through photosynthesis.

Some bacteria are producers as well. Most use a <u>process</u> called chemosynthesis (kee moh SIHN thuh sus). In chemosynthesis, energy from chemical reactions is used to make food. There is also one type of bacteria that makes food through photosynthesis. It is called cyanobacteria.

Academic Vocabulary

process (PRAH ses) (noun) a series of actions leading to a conclusion

Consumers

Organisms that cannot make their own food are called <u>consumers</u>. All animals, including humans, are consumers. This is because we must get energy by consuming other organisms or their wastes. We cannot carry out photosynthesis to create food ourselves. ✔

Some small consumers have only one cell. These animal-like protists are called <u>protozoans</u>. They feed on living or dead organisms. Protozoans have special structures to digest food and get rid of wastes.

✔ Reading Check

2. **Define** What are consumers?

_____

_____

_____

What are the types of consumers?

Think of the foods that animals eat. Different types of organisms get their energy from different types of food. Consumers can be classified by the kinds of foods they eat.

Herbivores Animals that eat only plants are **herbivores** (HUR buh vorz). Squirrels are herbivores. This is because they eat only plant products, such as nuts and seeds.

Carnivores **Carnivores** (KAR nuh vorz) are animals that only eat other animals. Some are very small. A spider is a carnivore. It lives by eating insects. A sea anemone is also a carnivore. It eats creatures that swim into the reach of its tentacles. Predators are animals that hunt and kill other organisms for food. The organisms they hunt and kill are called prey.

Omnivores Animals that feed on both animals and plants are called **omnivores** (AHM nih vorz). Grizzly bears are omnivores. They eat plant products such as nuts and seeds. They also eat animals, such as fish and elk. Most humans are omnivores as well.

What are scavengers and decomposers?

Scavengers (SKA vun jurz) feed on dead animals. Vultures and crows are scavengers. Some other species, such as foxes and coyotes, can be both predators and scavengers. They track live prey, but also feed on dead animals.

Decomposers (dee kum POH zurz) are organisms that break down wastes made by living things. These wastes can include leaves, dead organisms, animal droppings, and other items. Decomposers make nitrogen and other nutrients available to support new life. If they did not exist then these nutrients would remain in the bodies of dead organisms. This would limit the growth of new things. Many species of bacteria and fungi are decomposers. Some insects, protists, and other invertebrates are also decomposers. ☑

Organisms Depend on Each Other

In this lesson, you learned that producers, including green plants, some protists, and some bacteria use energy from the Sun to make their own food. Consumers gain energy from eating other organisms, including producers. Decomposers break down dead organic matter, making nutrients available for other organisms.

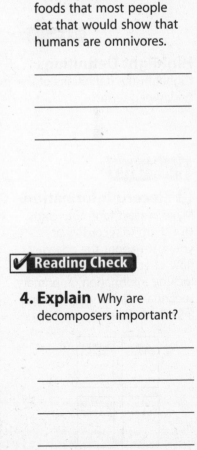

Think it Over

3. **Generalize** Name two foods that most people eat that would show that humans are omnivores.

_____

_____

_____

Reading Check

4. **Explain** Why are decomposers important?

_____

_____

_____

_____

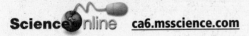

Energy and Matter in Ecosystems

lesson 2 Energy in Ecosystems

Grade Six Science Content Standard. 5.a. Students know energy entering ecosystems as sunlight is transferred by producers into chemical energy through photosynthesis and then from organism to organism through food webs. **Also covers:** 5.c.

MAIN Idea

Energy flows through ecosystems.

What You'll Learn

- how energy transfers from one organism to another
- about energy pyramids

Mark the Text

Highlight Definitions
Highlight the definitions of the underlined terms.

FOLDABLES

B **Record Information**
Make at least four note cards. Use them to record what you learn about the one-way flow of energy in ecosystems. Include information on primary, secondary, tertiary consumers, and food chains and webs.

Primary Consumers	Secondary Consumers
Tertiary Consumers	Food Chains and Webs

⬤ Before You Read

How does your body get energy? On the lines below, describe a few tasks you have already done today that required energy. Then read the lesson to learn how energy cycles throughout your world.

_____

_____

_____

_____

⬤ Read to Learn

Energy Through the Ecosystem

Every second that you are alive, you are using energy. You use energy to walk, talk, and even to breathe. Energy is the ability to do work. All living things need energy. However, energy does not cycle through ecosystems. Energy moves in one direction.

Energy starts in sunlight. It moves to producers, then to consumers, and then to decomposers. If producers stopped capturing energy from the Sun, all life on Earth would end. This is because food supplies would run out.

How do organisms change energy?

Organisms do not create energy. They also do not destroy energy. They do change energy from one form into another. During photosynthesis, producers change light energy into chemical energy. When consumers eat producers, some of the chemical energy in the food changes into heat. These are just two of the many ways in which organisms change energy into different forms.

Food as Energy

Where does your body get the energy it needs to walk, breathe, and do other things? It comes from the food you eat. The food you eat provides sugars, starches, proteins, and fats needed by your body to grow new cells. Food also provides your body with chemical energy your body uses as fuel. Energy enters the ecosystem when producers capture energy from sunlight. They use this energy to live and grow. Other organisms gain energy by eating producers. This gives them the energy that had been stored in producers' cells. In this way, the energy captured by producers becomes <u>available</u> to all living things.

What are food chains?

A <u>food chain</u> is a picture of one way that energy moves through an ecosystem. Most food chains start with the Sun. The next step in the food chain is a producer, such as a berry-filled bush shown below. Then there is a consumer that eats the berries from the bush, such as a mouse. The last step in this food chain is also a consumer, such as the black bear that eats the mouse. Notice that decomposers and scavengers are not usually shown in a food chain.

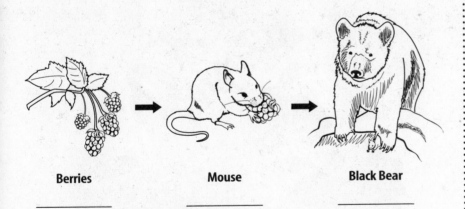

Berries   **Mouse**   **Black Bear**

_____   _____   _____

What are food webs?

Most organisms eat more than one type of food. A <u>food web</u> is a more complicated model of the way energy flows through an ecosystem. It is similar to a food chain in that it also begins with the Sun. However, food webs include many producers and many consumers. Food webs show that in ecosystems, energy flows through a large variety of species at the same time. ☑

Academic Vocabulary
available (uh VAY luh bul) (adj.)
present or ready to be used

Picture This

1. Identify Write *Producer* or *Consumer* below each organism on the food chain.

✔ **Reading Check**

2. Compare How is a food web similar to a food chain?

_____

_____

_____

_____

Think it Over

3. Apply What other pyramids are you familiar with?

_____

_____

Picture This

4. Predict Name three organisms that you have seen that belong on the first level of an energy pyramid.

_____

_____

_____

What are energy pyramids?

An energy pyramid is a picture that shows how much energy is available to different species in an ecosystem. Notice the pyramid below. The bottom level is very large. As you move toward the top, each level gets smaller. Each level of an energy pyramid contains the organisms described below.

- First level—Producers are found at the bottom of the pyramid. Notice in the energy pyramid below that this is the largest level.
- Second level—**Primary consumers** eat producers.
- Third level—**Secondary consumers** eat the primary consumers.
- Fourth level—**Tertiary consumers** are the top level of the energy pyramid. These predators eat organisms from the levels below.

How do producers use energy?

What happens to the energy captured by a producer? Some is used by the producer as it lives and grows. Some is lost as heat. The rest is stored in the producer's cells. Only the energy stored in its cells is available to the organism that eats the producer. For this reason, as you get farther up the energy pyramid, less and less energy is available. ☑

What determines the size of an energy pyramid?

The number of producers in a biome will determine how large the energy pyramid will be. There are more producers in a rain forest than in a desert. The energy pyramid for the rain forest biome will have a larger first level than the energy pyramid for the desert. Because the first level is larger, the entire energy pyramid for a rainforest is larger than the energy pyramid of the desert biome.

What do cars and organisms have in common?

A car gets its power from the energy stored in gasoline. Not all of the energy in gasoline is burned in its engine to make the car move. Much of the energy is released to the atmosphere as heat. Once released, heat energy cannot be recaptured. The car must be refilled with gas when the gas tank is empty.

In ecosystems, producers use energy from the Sun to make food. As long as the Sun shines and producers are present, food will be made for life on Earth. Primary consumers eat producers for energy and nutrients. However, not all of the energy stored in the producer is available for the activities and growth of the consumer. Like the car engine, much of the energy is released as heat. Tertiary consumers have much less energy available to them than primary consumers. This is why energy pyramids are larger at the bottom than they are at the top.

✔ **Reading Check**

5. **Explain** Why is less energy available farther up the energy pyramid?

_____

_____

_____

💡 **Think it Over**

6. **Identify** On which level of the energy pyramid are human beings?

_____

Energy and Matter in Ecosystems

lesson ❸ Matter in Ecosystems

 Grade Six Science Content Standard. 5.b. Students know matter is transferred over time from one organism to others in the food web and between organisms and the physical environment.

MAIN Idea

Matter cycles in ecosystems.

What You'll Learn

- the cycles of matter
- where matter comes from for tree growth

> Study Coach ▶

Summarize Main Ideas

As you read, write one sentence to summarize the main idea in each paragraph. Use underlined words in your sentences.

Picture This

1. **Highlight** the arrow to illustrate the cycle.

● Before You Read

List three things your body needs to survive. Then describe where those things come from. Read the lesson to learn about how matter moves in various cycles.

_____

_____

● Read to Learn

Cycles of Matter

Living and nonliving things are made of matter. The total matter on Earth never changes. Instead, matter changes from one form into another. For example, when a tree grows, some matter from the environment becomes part of the tree's cells. It is changed into living parts of the tree.

Water Cycle

One important type of matter is water. Earth has only a certain amount of water. The same water is used repeatedly as it cycles through the environment as shown below. It evaporates from lakes and oceans, returns to the earth as rain or snow, and is used by plants or animals. When plants die or release water in their waste products, that water enters the atmosphere. It begins cycling through the environment again.

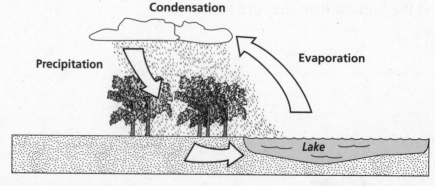

Nitrogen Cycle

The cells of all living things need nitrogen. Nitrogen is plentiful in the air. However, most living things cannot use nitrogen in this form. It is changed into a form that plants can use by a type of soil bacteria, called **nitrifying bacteria**.

After nitrogen is changed into a new form, plants can take it up through their roots. It becomes part of their cells. If an animal eats that plant, the nitrogen becomes part of the animal's cells. In this way, nitrogen enters the food chain and is used by all living things.

Nitrogen returns to the environment when organisms decay. It is always cycling through living organisms and their environment as shown below. The **nitrogen cycle** describes how nitrogen moves from the atmosphere to the soil, to living organisms, and back to the atmosphere.

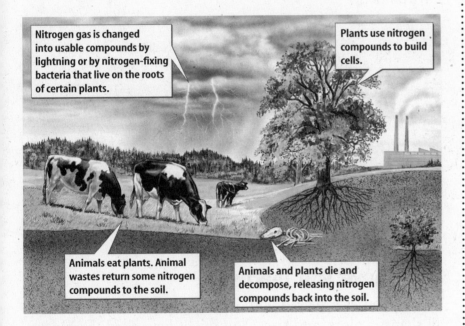

Nitrogen gas is changed into usable compounds by lightning or by nitrogen-fixing bacteria that live on the roots of certain plants.

Plants use nitrogen compounds to build cells.

Animals eat plants. Animal wastes return some nitrogen compounds to the soil.

Animals and plants die and decompose, releasing nitrogen compounds back into the soil.

Phosphorous Cycle

Phosphorus is <u>found</u> in the soil. It gets there from the weathering of rocks. Phosphorus is a nutrient that plants take up through their roots.

The **phosphorus cycle** describes how phosphorus moves from soil to producers, to consumers, and back to soil. Animals get phosphorous by eating plants. They can also get it by eating other animals that have eaten plants. Phosphorus returns to the soil when dead plants and animals decompose, and also through the waste products of animals.

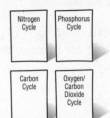

FOLDABLES™

C Record Information

Make at least four note cards. Use them to record what you learn about the nitrogen, phosphorous, carbon, and oxygen/carbon dioxide cycles.

Nitrogen Cycle

Phosphorus Cycle

Carbon Cycle

Oxygen/ Carbon Dioxide Cycle

Picture This

2. Explain How does nitrogen become part of human cells?

_____

_____

_____

_____

Academic Vocabulary

found (FOWND) (verb) located or discovered

The Carbon Cycle

The **carbon cycle** describes how carbon moves between the living and nonliving environment. Carbon is an essential nutrient for all organisms. This is because carbon is found in the sugars, proteins, starches, and other compounds that make up living cells. Living things must have carbon in order to grow. ☑

How does carbon move between the living and nonliving environment?

Carbon dioxide is taken from the air during photosynthesis. Notice below how and where photosynthesis brings carbon into the food chain. It is sent back into the air by all organisms during a process called respiration. In this way, carbon moves between the living and nonliving environment.

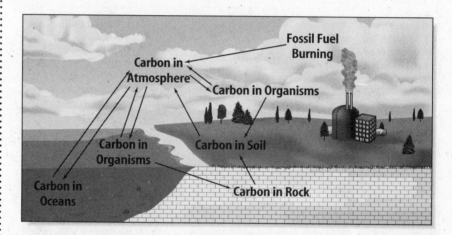

3. Name three parts of a cell that are made with carbon.

_____

_____

_____

Picture This

4. Identify Circle the area where burning fossil fuels adds carbon dioxide to the atmosphere.

What have you learned?

Have you ever played with building blocks? They come in all shapes, colors, and sizes. You might have used the blocks to make buildings, or even a whole town. What happened when you had finished your building? Perhaps you tore it down and used the same blocks to build a spaceship or a car. Like building blocks, matter on Earth is used to build organisms. When organisms die, decomposers tear down the matter into their basic building blocks.

It took energy for you to use the individual building blocks to make a castle, car, or boat. In the same way, organisms, including people, need energy for growth and daily activities.

Science Online   ca6.msscience.com

Resources

chapter
14

lesson ❶ Natural Resources

Grade Six Science Content Standard. 6.b. Students know different natural energy and material resources, including air, soil, rocks, minerals, petroleum, fresh water, wildlife, and forests, and know how to classify them as renewable or nonrenewable.

● Before You Read

What do you think of when you hear the word resources? On the lines below, write your ideas about the resources you need to get ready for school each day. Read the lesson to learn about the variety of material resources found on Earth.

_____

_____

● Read to Learn

Organic Resources

<u>Natural resources</u> are materials and energy sources that are useful or necessary to meet the needs of Earth's organisms, including people. Plants and animals that are living, or recently were alive, are organic material resources. Think about your food. Nearly all of it is either animal or plant material. Some clothes, such as denim jeans, are made from cotton plants. Wool, silk, and leather are materials from animals. Many homes are made of wood and most home furnishings are manufactured from plant materials. All of these are made from organic material resources.

Inorganic Resources

Not all natural resources come from plants or animals. Inorganic material resources include metals and minerals, which did not come from living organisms. In Ghana, Africa, gold, aluminum, diamonds, and manganese are mined and exported to other countries. Gold and diamonds are used in jewelry and electronics. Aluminum is used to make soda cans, bicycles, and other common items that need to be strong and lightweight. ☑

MAIN ⟨Idea

People use a variety of materials from different parts of Earth to meet a diverse range of needs.

What You'll Learn

■ material resources used to make common objects
■ Earth's material resources as renewable or nonrenewable

▸ **Study Coach**

Identify the Main Point
Highlight the main point in each paragraph. Then use a different color to highlight a detail or example that helps explain each main point.

☑ **Reading Check**

1. **Identify** Which of the following are inorganic material resources? (Circle your answer.)
   a. metals and minerals
   b. plants and animals

What are some uses of manganese?

Manganese is a silver-colored metal found in rocks, usually combined with oxygen or sulfur. Manganese must be separated from other elements in rocks. Manganese is used to make pesticides and is added to gasoline to help cars run better.

How does the cost of mining relate to the value of a mineral?

Many minerals like gold, silver, and manganese must be mined, or dug out of the ground. A large amount of a mineral must be in one area of rocks to make it worth the cost of the mining operation. Small amounts of minerals in rocks may not be worth mining. The cost to separate the metal or minerals from the earth may be more than what the minerals are worth. The value of minerals changes and so does the cost of the equipment to find and take out the resources. Sometimes the price for a mineral is high. Mining companies can spend more money on the process of getting the minerals out of the ground when the price is high. ✔

How does the building industry use inorganic resources?

Other inorganic resources include industrial and building materials. Steel is made from iron that is taken from the Earth. Sand and gravel are very valuable and important for construction. Sand is used to make concrete for buildings, sidewalks, and bridges. The sand and gravel production in California is worth more than $1 billion each year. California makes more money mining sand than gold.

Renewable Resources

Earth's resources that are being replaced by nature are called **renewable natural resources**. You might think that all organic resources are renewable because they are made from living things. But organic resources are only renewable if we manage them carefully. For example, Douglas fir trees in Oregon are used for lumber. These trees take 200 years old to grow into large tress. It might be difficult to replace Douglas fir trees if many are cut down in a short time. Some trees grow faster and the populations of these trees are easier to manage.

---

2. Explain When is it worth the cost to extract minerals?

_____

_____

_____

_____

FOLDABLES™

A Explain Make a two-tab Foldable. Label the tabs as illustrated. Under the tabs, explain renewable and nonrenewable natural resources in your own words and give examples of each.

Renewable Natural Resources

Nonrenewable Natural Resources

How is the United States managing forests?

Careful management of forests is important for the environment. Vegetation, like trees, give off oxygen and remove carbon dioxide from the air. Trees also slow erosion, provide homes for wildlife, clean pollutants out of the air, and filter rain water that runs off land to streams.

In the United States, 155 forests are managed by the U.S. Forest Service, including many forests in California. The U.S. Forest Service determines the number of trees that can be cleared, or cut down, and the number of trees that must be planted.

Think about other <u>resources</u> that may be affected when cutting trees. Roads must be built to allow heavy machines into forested areas to cut trees. Road building can destroy forest habitat. Although young trees usually are planted to replace those that are lost, the forest habitat may not be completely replaced. In this case, the forest habitat is a nonrenewable resource. ☑

How do habitats affect animal resources?

When fish are caught for food faster than the fish can reproduce, this valuable renewable resource is affected. But fish populations also can decrease if areas they need to grow and reproduce are destroyed. An <u>estuary</u> is a fertile area where a river meets an ocean. Estuaries have a mixture of freshwater and salt water. Estuaries serve as nursery areas for many types of fish and other organisms. But the ocean and its harbors are important resources for humans, too. Goods are transported by ships into harbors. The beautiful ocean view makes coastal areas ideal for marinas, houses, and hotels.

In the last two hundred years, the area occupied by wetlands in the United States has decreased by almost half. In California and many other areas of the United States, nearly 90 percent of coastal wetlands have been developed. The numbers of young fish that hatch each year and grow into adults will continue to decrease if their nursery areas are destroyed.

Populations of many organisms in California and the United States are getting smaller and smaller. California has nearly 300 threatened and endangered plant and animal species. Plants and animals are interconnected through food webs. A decrease in the population of one species may have a bad effect on species that serve as resources for people. ☑

Academic Vocabulary

resource (REE sors) (noun) a natural source of materials

☑ **Reading Check**

3. **Describe** What are the benefits to people and animals of managing our forests well?

_____

_____

_____

_____

_____

_____

_____

☑ **Reading Check**

4. **Identify** What state has nearly 300 threatened and endangered species?

_____

Nonrenewable Resources

Resources that are used more quickly than they can be replaced by natural processes are called **nonrenewable natural resources**. These resources are used at rates far faster than their geologically slow formation rates. Gold, a nonrenewable resource, is deposited when hot water and molten rock, called magma, flows through spaces in underground rock. The hot magma heats water and gold travels with mineral solutions in the water. When the magma and solution cools, gold collects. ☑

Why is gold a nonrenewable resource?

Gold is extracted from two types of mines in California. Gold is a nonrenewable natural resource because it is removed from Earth faster than it is created. Gold is worth a lot of money because it can be formed into various shapes. Gold is pleasing to look at but there is a limited amount of it on Earth.

What was the California Gold Rush?

In 1848, large veins of gold were discovered in California. People traveled to California in hopes of striking it rich. The people who came to California were called "49ers" because most of the left home in 1849 to travel to California. Many people traveled from the eastern part of the United States. Other came from countries around the world seeking a fortune.

At first, gold was easy to find. But in a short time, it became hard to make money because less and less gold was found in the mines. Those who did find gold spent most of their money on the supplies they needed to live. The people who made the most money during the California Gold Rush were merchants who sold food, shovels, clothing, and other goods.

What kind of water is a nonrenewable resource?

All the water on Earth is already here. Currently, there is no way to create new water. Freshwater is an important nonrenewable resource for California. Even though most of California is arid and dry, people use large amounts of water for irrigation, industry, and personal use. Think about the size of a large, 1-L soft-drink container. How many liters of water do you think you use each day? The table on the next page lists average water uses in the United States.

✔ Reading Check

5. Explain why gold is considered a nonrenewable resource.

_____

_____

_____

💡 Think it Over

6. Draw Conclusions Why did merchants make money during the California Gold Rush?

_____

_____

_____

Average Daily Water Use for a Family	
Daily Activity	Water Used
Flushing the toilet once	15 L
Taking a short shower	95 L
Taking a longer shower	190 L
Taking a bath	150 L
Washing clothes	190 L
Automatic dishwasher	38 L
Brushing your teeth while leaving the water running	7.5 L
Washing your hands while leaving the water running	30 L/min
Watering the lawn or plants with a hose	30 L/min

Picture This

7. Estimate Using the table, estimate how much water your family uses every day.

Where does California get its supply of freshwater?

California gets some of its freshwater from the Colorado River. Water from the lower Colorado River is divided between four states and the American Indian tribes. Many laws and treaties with Mexico have been passed to try to regulate the use of this water resource. As other states in the Colorado basin have increased their use of the river's water, Californians are trying to reduce the amount of water used each day. One way to solve the problem of limited water supplies is to reuse water. Water that has been reclaimed from city wastewater can be treated and reused. ☑

What have you learned?

Earth has large amounts of natural resources. Humans have learned to use these resources to make materials for everyday use. Renewable resources can be replaced as fast as they are used. Other resources are used faster than they can be replaced. These resources are nonrenewable.

In the next lesson you will read about how some of Earth's resources are used for energy. As you read, think about ways that you can conserve natural resources by using less energy.

✔ **Reading Check**

8. Name one way that Californians can conserve water.

_____

_____

_____

Science nline   ca6.msscience.com

chapter 14

Resources

lesson ❷ Energy Resources

Grade Six Science Content Standard. 6.a. Students know the utility of energy sources is determined by factors that are involved in converting these sources to useful forms and the consequences of the conversion process. **Also covers:** 6.b.

Some of Earth's natural resources can be used for energy.

What You'll Learn

- the development and extraction of fossil fuels
- the energy resources that eventually are used to create steam to turn turbines
- the advantages and disadvantages of each energy resource

Study Coach

Make A Sketch As you read, draw your own sketches to help you understand and remember new information.

FOLDABLES

Ⓑ Define and Record
Make a four-door Foldable. Label the front tabs as illustrated. Under the tabs, define and record information you learn about energy resources. Explain the past, present, and future importance of fossil fuels.

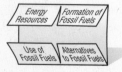

● Before You Read

Flashlights are very helpful for going on a walk in the dark. Think about how long the batteries can last. How many flashlights and batteries would it take to light up your home at night. Write your ideas on the lines below. Then read the lesson to learn more about Earth's energy resources.

_____

_____

_____

● Read to Learn

Fossil Fuels

<u>Fossil fuels</u> are fuels formed in Earth's crust over hundreds of millions of years. Coal, oil, and natural gas are fossil fuels that supply energy. Most of the energy you use comes from these fossil fuels. Cars, buses, trains, and airplanes are powered by gasoline, diesel fuel, and jet fuel. All these fuels are made from oil. Coal is used in many power plants to produce electricity. Natural gas is often used in manufacturing and for heating and cooking. People on Earth are using more fossil fuels for energy today than ever before.

How do we get energy from fossil fuels?

The best energy sources can be easily changed to heat, electricity, and used for transportation. Today, people use fossil fuels as the main source for energy because we can make energy from fossil fuels into forms we can use.

Fossil fuels contain a lot of stored energy and burn easily. For example, fossil fuels are burned to heat water and make steam. The steam then is used to turn large turbines that power generators, which create electricity.

Formation of Fossil Fuels

Fossil fuels are made from decayed plants and animals. The plant and animal material has been changed into fossil fuels by pressure, bacterial processes, and heat. About 300 million years ago, before the time of the dinosaurs, Earth was covered with green, leafy plants. Large amounts of algae and other small organisms grew in Earth's oceans, lakes, and rivers. Over many hundreds of years, the decaying organisms were covered by sand and clay. The sand and clay layers formed into rock. ✔

More and more layers of rock were formed over the decaying plants. The weight of the rock layers pressed down on the decaying material. Eventually, over millions of years, heat and pressure from layers of rock pressing down turned the plant and animal remains into the three main forms of fossil fuels—coal, oil, and natural gas. The formation of each fuel is described in the table below. Fossil fuels are nonrenewable resources because they cannot form as fast as they are used.

Formation of Fossil Fuels

Fossil Fuel	Description
Oil	1. Microscopic plants and bacteria are the main source of oil. Some of these organisms were producers, using energy from the Sun to make food for growth and reproduction. When the organisms died, they fell to the seafloor. 2. The microscopic organisms were buried under clay. 3. Many layers of clay and mud increased the pressure and temperature, forming liquid oil.
Coal	1. Coal is the most abundant of all the fossil fuels. 2. Coal formed from the incomplete decay of plants. The partially decayed plant material, called peat, becomes sandwiched between layers of sediment. 3. Soft coal forms under moderate pressure and heat. 4. As more heat and pressure are applied, the soft coal becomes hard coal.
Natural gas	1. Natural gas forms along with oil. Because it is less dense than oil, natural gas is usually found above oil deposits. 2. Natural gas is usually associated with oil in deposits that are 1 to 3 km below Earth's crust. 3. Natural gas is found in areas beneath layers of solid rock. The rock prevents the gas from escaping to the surface.

1. Explain What are fossil fuels made of?

_____

_____

_____

Picture This

2. Identify What is decayed plant material called? (Circle the answer.)

a. oil

b. peat

How did oil form?

Oil or petroleum is used for heating. It can be refined into gasoline, kerosene, or diesel fuel. As you will read later in another lesson, petroleum also is used for many other materials you use every day.

Oil formed mainly from ancient, microscopic plants and bacteria that lived in the ocean and salt water seas. When these organisms died they were quickly buried by clay. The remains did not completely decay but formed mud rich in materials that form fossil fuels. The temperature and pressure increased as more clay and mud collected on top of the dead organisms. Liquid oil formed as temperature and pressure increased. The oil flowed into spaces in the rock, shown below.

To get the oil out of the rock spaces, geologists drill wells deep into Earth. Most of this oil is under pressure and flows to the surface where it is collected. Oil collected out of the ground is sometimes called crude oil. Crude oil must be refined for use in cars, trucks, airplanes, and trains.

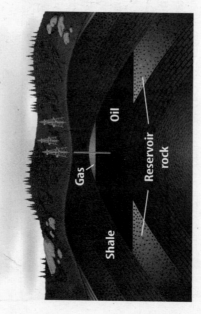

Gas

Oil

Shale

Reservoir rock

What is natural gas?

Natural gas forms along with oil. Natural gas is less dense and so it is found on top of oil pools. Natural gas is valuable because it burns cleanly and can be moved easily in underground pipelines. Some families and businesses use natural gas for heating their buildings and cooking food. ✓

How is coal used?

There is more coal than any other fossil fuel in the world. Coal forms where vegetable matter collects but is prevented from complete decay. The partially decayed plant material forms layers of spongy material called peat. Over time the peat becomes buried and compressed. Increased pressure and temperature produce a solid, sedimentary rock, known as coal.

Academic Vocabulary

refine (ree FINE) (verb) to improve; to clean or remove impurities

Picture This

3. Identify Circle the areas where oil and natural gas are found.

✓ Reading Check

4. Name two reasons that natural gas is valuable.

_____

_____

_____

How did coal develop?

Coal formed under moderate amounts of heat and pressure is called soft coal. If more heat and pressure is applied to soft coal, it can become hard coal. Soft coal contains some moisture and sulfur. When soft coal burns, it can release pollutants into the air. Hard coal burns more cleanly. It has the greatest amount of carbon, so it provides more energy. Coal is removed from the earth by strip mining and underground mining.

How is strip mining done?

In strip mining, as shown below, machinery is used to scrape the plants, soil, and rock layers off the ground above a vein of coal. Machines remove the surface soil, trees, and other organic material, and place it beside the mine. Once the layer of coal is exposed, it can be removed and loaded into containers for transportation. Strip mining is less expensive than underground mining if the coal is close to the surface.

How else is coal removed from the earth?

If the coal is very deep underground mines are used. Underground mines are created by digging down into the earth at an angle to form tunnels. These tunnels go deeper and deeper until they reach the coal. Wooden beams and pillars support the tunnel and make it safer for the miners. In some mines, tracks are laid to allow wagons to roll along the tunnels to move the coal toward the surface.

Mining coal produces a fine, black coal dust. If miners inhale the dust, their lungs can be damaged. In the past, many miners suffered from a disease called black lung. Today, miners wear protective clothing and masks that keep them from inhaling the coal dust. ☑

Picture This

5. Predict What problems might result from this strip mine?

_____

_____

_____

✔ **Reading Check**

6. Explain Why do underground coal miners need to wear masks?

_____

_____

_____

Alternatives to Fossil Fuels

Energy doesn't have to come from burning fossil fuels. Alternative energy sources, including water, wind, ocean waves, and natural heat sources beneath Earth's surface can be used to produce electricity.

What is hydroelectric power?

Hydroelectric power is a renewable resource. As illustrated in the figure below, large dams block the flow of water from major rivers and create lakes behind the dams. As the water moves rapidly through the narrow openings in the dam, turbines generate electricity. The amount of water that moves through the turbine can be increased to generate greater amounts of electricity.

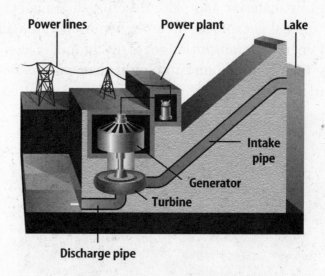

Power lines   Power plant   Lake

Intake pipe

Generator

Turbine

Discharge pipe

How is wind energy harnessed?

Energy can be produced from wind. Scientists are looking for less costly ways to use wind to produce energy. Near Palm Springs, California, wind farms with long rows of wind towers connect to generators. Wind towers need a steady wind that is not too strong or too weak. Wind farms create no pollution and have been successful on a small scale. Because winds vary in speed, wind energy will likely be used along with other forms of energy. ☑

What is geothermal energy?

The heat energy in Earth's crust is called **geothermal energy**. The extreme heat found in rocks near geysers and volcanoes can be used to generate steam for electricity. Geothermal energy is clean and safe, but there are only a few places where enough heat is near the surface.

Picture This

7. Interpret Diagrams
Place arrows on the figure to show the direction of water flow through the hydroelectric power plant.

✔ **Reading Check**

8. Explain What is a problem with wind towers?

_____

_____

_____

How is nuclear energy made?

All matter is made of atoms—tiny particles that we cannot see. Even though atoms are very small, when they split, a large amount of energy is released. Splitting atoms to release energy is called **nuclear fission**. Atoms from the element, uranium, are split in a nuclear reactor. When atoms split, the energy that is released heats water in the reactor. Steam is produced and turns a turbine. The turbine runs a generator that creates electricity. ☑

Combining atoms also makes heat. In a **nuclear fusion** reaction, as shown below, two atoms of the element, deuterium, join together to form one atom. It's the same type of reaction that powers the Sun. Like nuclear fission, nuclear fusion gives off large amounts of energy. Fusion reactions are not easy to start. In order to get the atoms to fuse and start the fusion reaction, temperatures must be over 100,000,000°C.

Earth has a lot of uranium and very little of it is needed to supply a nuclear fission reactor. Deuterium can be made from water. Earth has a plentiful supply of this potential fuel.

✔ **Reading Check**

9. **Define** What is nuclear fission?

_____

_____

_____

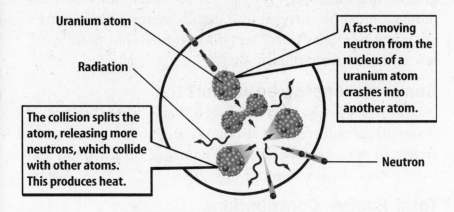

Uranium atom

Radiation

A fast-moving neutron from the nucleus of a uranium atom crashes into another atom.

The collision splits the atom, releasing more neutrons, which collide with other atoms. This produces heat.

Neutron

Picture This
10. **Describe** Use the figure to explain to a partner how heat is produced from uranium.

How is solar energy captured?

Sunlight is a never-ending resource because we can never use up all the energy from the Sun. Solar energy is the process of changing the light and heat energy from the Sun into electricity. Solar cells are used to change energy for calculators and other small appliances. Solar panels are made up of many solar cells that capture energy from the Sun. The energy is stored in a series of batteries for later use. Large amounts of solar energy are needed to heat larger buildings or run large appliances. Scientists are trying to find a way to collect and save a lot of energy from the Sun.

Solar-Generated Power Scientists in California are experimenting with solar towers to make energy. Panels are constructed at the bottom of large towers. The panels collect sunlight and heat up. As the warm air rises, it turns turbines that generate electricity. Solar energy could be a clean, endless supply of energy in the future.

Problems with the Use of Solar Power Solar energy equipment is expensive and the batteries store only a small amount of energy. Another problem is that solar panels do not work at night or on very cloudy days.

Solar energy is not practical to power vehicles such as cars and trucks. Large batteries are needed to store the energy. The batteries add to the weight of the vehicle. Vehicles that weigh more use more energy. In addition, batteries can take up too much space in a small car.

How can biomass be used for energy?

Organic matter that makes up plants is known as biomass. Biomass can be used to produce fuels for electricity and transportation. Food crops, such as corn and soybeans, grasses, trees, and even garbage are forms of biomass. Most biomass must be changed into usable energy forms. There are many refineries to convert oil into fuels, but there are few refineries for biomass.

Can wave energy be useful?

Using the energy from waves is difficult. There are not many places with regular, strong wave action, where a system of turbines would not be damaged on the rocks and would work during low and high tides.

Total Energy Contributions

In the United States, about 40 percent of energy use is from oil, 23 percent is from coal, 22 percent from natural gas, eight percent from nuclear energy, and three percent from hydroelectric power. The rest is from a combination of geothermal, solar, and wind energy. ☑

What have you learned?

Billions of people all over the world use fossil fuels every day. Fossil fuels are nonrenewable; Earth's supply of them is limited. In the future, fossil fuels may become more expensive and difficult to obtain. Developing other forms of energy can reduce pollution and conserve resources.

💡 Think it Over

11. Summarize What are some of the reasons that solar power is not the perfect source for energy yet?

_____

_____

_____

_____

_____

_____

✔ Reading Check

12. Calculate What percent of the total energy use is the combined energy from geothermal, solar, and wind energy?

_____

Science Online   ca6.msscience.com

chapter 14 Resources

lesson ❸ Using Energy Resources

Grade Six Science Content Standard. 6.c. Students know the natural origin of the materials used to make common objects. **Also covers:** 6.a.

● Before You Read

Imagine there are no grocery stores and no department stores. How would you make clothes and food? Write your ideas below. Read the lesson to learn about the energy resources of the United States.

_____

_____

_____

● Read to Learn

Location of Natural Resources

Different resources are found in different parts of the world. If the resource is hard to find or collect it may be expensive. Buying and selling resources is an important part of the economy for different areas of the United States and other countries. The table below shows the location of some resources in the United States.

Resources from Various States	
Alabama	cement, limestone, cotton, lumber
Alaska	oil, fish, lumber, zinc, gold, sand and gravel
Arizona	cotton, copper, sand and gravel
California	milk, grapes, flowers, sand and gravel, cement, boron
Delaware	shellfish, soybeans, sand and gravel
Florida	oranges and lemons, crushed stone, fish
Hawaii	sugar, pineapple, nuts, fish
Minnesota	lumber, iron, corn
Montana	lumber, gold, coal, natural gas
Texas	oil, natural gas, crushed stone
Washington	lumber, sand and gravel, apples, fish

MAIN Idea

Conserving resources can help prevent shortages and reduce pollution.

What You'll Learn

■ common objects made from natural resources
■ strategies to conserve energy

Study Coach

Summarize As you read this lesson, stop after each paragraph and summarize the main idea in your own words.

Picture This

1. List the nonrenewable resources found in the table.

Manufacturing Common Objects

Fuel for energy is not the only product from fossil fuels. Oil is used to manufacture many common materials, including most plastics and synthetic clothing, such as nylon. When you use a plastic fork or open a plastic package, you are holding objects made from fossil fuels. Reusing plastics and passing on your outgrown clothes to others will help extend our energy resources. ☑

For most products you use, packaging the objects, transporting them, keeping the stores a comfortable temperature for shopping, and the other steps in making and selling things use Earth's material resources and energy.

What is recycling?

People can decrease their use of natural resources by reducing the amount of products they purchase, choosing products that do not use too much packaging, reusing products, and recycling. **Recycling** is changing or reprocessing an item or natural resource for reuse. Materials such as plastics and aluminum can be recycled. Recycling uses less energy than extracting new natural resources and helps natural material and energy resources last longer.

Drawbacks of Using Fossil Fuels

There are many drawbacks to fossil fuel use. In order to refine oil into gasoline and other fuels, the crude oil must be transported from the drilling site to the refinery. Ships are the most economical way to transport crude oil. Marine life can be greatly affected if crude oil spills from a ship.

Burning fossil fuels releases carbon dioxide into the air. Carbon dioxide contributes to global warming. Burning coal gives off sulfur dioxide. This gas causes acid rain. When vehicles burn gas, they release pollutants that contribute to the smog in many cities. Burning fossil fuels also releases tiny particles, called **particulates**, into the air. Particulates can damage lungs.

How is the amount of pollution controlled?

Manufacturing products from oil can also pollute our air and water. National and state laws and regulations tell companies how much pollution they are allowed to release into the environment. Local, state, and national government policies and economics influence these regulations.

2. Identify What materials, other than fuels, are made from oil?

_____

_____

FOLDABLES

⊙ Explain Make a two-tab Foldable. Label the tabs as illustrated. Under the tabs, explain how the use of fossil fuel affects the environment and give examples of alternative energy sources.

Fossil Fuel and the Environment

Alternative Energy Sources and the Environment

What is the impact of driving cars?

Americans use about one quarter of the world's oil. Most of the oil is used to power vehicles. There are about 24 million vehicles registered in the state of California alone. The average vehicle uses one gallon of gas for every 18 miles it travels. If Californians drove vehicles that averaged 30 miles for every gallon of gas, they would save more than 500 million gallons of gasoline each month.

Driving smaller cars can save gasoline. Other ways to conserve gasoline include carpooling, bicycling, walking short distances, and using buses or trains instead of driving. Scientists are also developing electric hybrid technologies that burn less fossil fuels. They are also researching ways to use fuels other than gasoline to power cars.

What is the impact of using coal?

About 70 percent of the electricity produced in the United States is generated using fossil fuels, especially coal. Coal must be extracted from Earth, transported, processed, and burned to generate electricity. Strip mining damages the land. Landslides, erosion, and polluted streams and lakes often result from strip mining. The land is generally useless after the mining ends, unless great care is taken to reclaim the land.

When coal burns, gases are released. Some gases help to form acid rain, which can damage forests, lakes, and streams. Burning coal also releases particulates and some toxic metals into the air. ☑

How does alternative energy use affect the environment?

Burning fossil fuels can cause air pollution. Extracting and transporting fossil fuels can damage land and water resources. Alternative energy sources can also harm the environment.

Although hydroelectric power is a renewable resource, large lakes that are created by dams block the flow of rivers and destroy wildlife habitats. Dams also increase the amount of sediment and erosion downstream. Dam failure can cause devastating floods.

Nuclear fission power plants produce dangerous radioactive waste. The waste must be stored safely for 10,000 years. Nuclear fission does not pollute the air with gases. However, a nuclear reactor <u>generates</u> a large amount of heat. Like all engines, nuclear power plants must get rid of unusable heat energy. Water is used to cool a nuclear power plant. Heated water released into the environment can harm fish and other wildlife.

3. Calculate Suppose your family is planning to take a 180-mile trip. What is the difference in number of gallons use in taking a car that gets 18 miles to the gallon or one that gets 30 miles to the gallon? Show your work.

_____

_____

_____

✓ Reading Check

4. Identify How does burning coal damage forests, lakes, and streams?

_____

_____

_____

Academic Vocabulary
generate (jeh nuh RAYT) (verb) to produce or create

Wind Farms Wind towers do not heat water or pollute the air. Wind farms do not destroy the environment permanently, but they do stop birds from using the area for nesting or feeding habitats. The best place to locate wind farms may be near large bodies of water, where it is windy. People might object to large wind towers obstructing their view. ☑

✔ **Reading Check**

5. **Compare** Name one advantage and one disadvantage to wind as a power source.

_____

_____

_____

Using Energy Resources Wisely

If people continue to use nonrenewable energy resources at current levels, as shown below, there could be shortages in the future. Few people want to completely change their lives to avoid using nonrenewable energy resources, but there are many ways to reduce the amount of resources all of us use each day.

Conservation means the preservation and careful management of the environment, including natural resources. Conserving nonrenewable resources is one of the most effective ways to help prevent shortages.

Picture This

6. **Estimate** How much electricity would you estimate your family uses on an average day?

_____

_____

_____

_____

Average Amount of Electricity Used by Common Household Appliances			
Appliance	**Average Electricity Used per Hour**	**Appliance**	**Average Electricity Used per Hour**
Lightbulb	100	Oven	1,300
Stereo	100	Air conditioner	1,500
Television	230	Hair dryer	1,500
Washing machine	250	Microwave	1,500
Vacuum cleaner	750	Clothes dryer	4,000
Dishwasher	1,000	Freezer	5,100
Toaster	1,200	Refrigerator/freezer	6,000

What have you learned?

Earth's many resources supply materials and energy for all of Earth's organisms. People use these resources for everyday living. Humans use fossil fuels as their primary source of energy. Developing other energy sources will reduce demand for fossil fuels and prevent pollution. Conservation and recycling can help save resources and reduce pollution.

Science Online   ca6.msscience.com

PERIODIC TABLE OF THE ELEMENTS

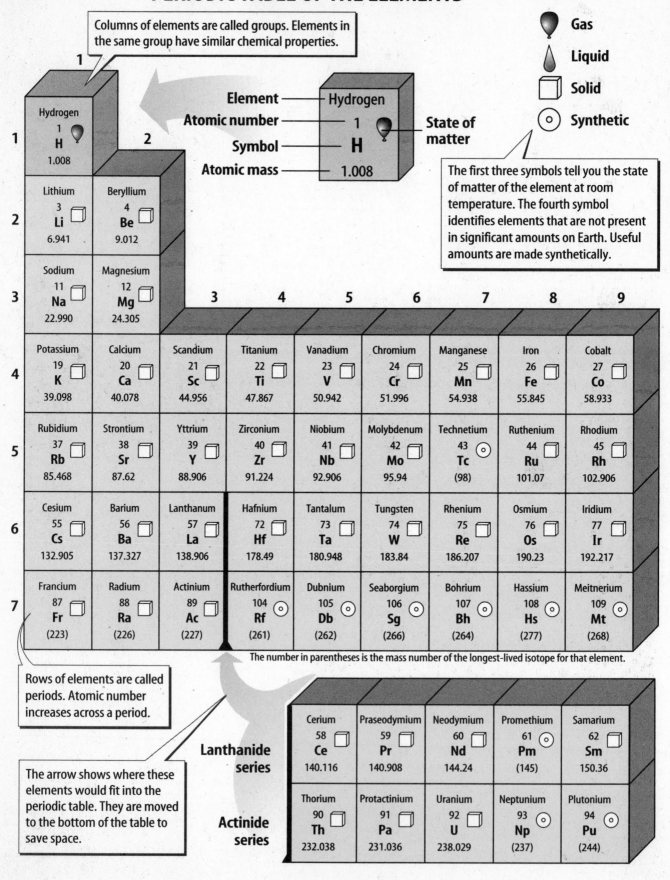

Columns of elements are called groups. Elements in the same group have similar chemical properties.

Gas
Liquid
Solid
Synthetic

Element — Hydrogen
Atomic number — 1
Symbol — H
Atomic mass — 1.008

State of matter

The first three symbols tell you the state of matter of the element at room temperature. The fourth symbol identifies elements that are not present in significant amounts on Earth. Useful amounts are made synthetically.

The number in parentheses is the mass number of the longest-lived isotope for that element.

Rows of elements are called periods. Atomic number increases across a period.

The arrow shows where these elements would fit into the periodic table. They are moved to the bottom of the table to save space.

Lanthanide series

Actinide series